Tiny Earth

A Research Guide to Studentsourcing Antibiotic Discovery

Simon Hernandez
Tiffany Tsang
Carol Bascom-Slack
Nichole Broderick
Jo Handelsman

XanEdu

Printed in the United States of America

ISBN 13: 978-1-71147-072-6

Cover Image Credit: Dr. Victor J. Cid of the Complutense University of Madrid

XanEdu
17177 Laurel Park Drive
Suite 233
Livonia, MI 48152
800-562-2147
www.xanedu.com

Contents

Acknowledgments

Thank you to our partner instructors (our TEPIs) for their thoughtful feedback and helpful suggestions. Special thanks to the hard work of Todd Kelson of Brigham Young University, Sachie Etherington from University of Hawaii, Michael Buckholt from Worcester Polytechnic University, Erica Suchman from Colorado State University, Barbara Murdoch from Eastern Connecticut State University, Ann Buchmann from Chadron State College, Ana Barral from National University, Kristen Butela from Seton Hill University, Jean Schmidt from University of Pittsburgh, Karen Pelletreau from University of Connecticut, Mutafa Morsy from University of West Alabama, and Neil Enis from Tulsa Community College.

Also, thanks to Sam Rikkers, **Tiny Earth** Executive Director, and Patricia Pointer, Multimedia Coordinator for the Wisconsin Institute for Discovery, for their help in producing these materials.

Finally, thanks to all our **Tiny Earthlings** across the world for sharing the journey of discovery and learning.

Introduction

Tiny Earth is a global network of students and instructors conducting research on soil and antibiotics. The network has four goals:

1. Educate students about the antibiotic crisis.
2. Educate students about the soil crisis.
3. Engage students in original scientific research.
4. Discover new antibiotics from soil!

Goal 1. The Antibiotic Crisis

There is a crisis brewing with potential for enormous health, economic, and political consequences. Many diseases caused by bacteria are becoming untreatable due to the proliferation of pathogens resistant to the effects of antibiotics. The World Health Organization (WHO) reports that, in the European Union, about 25,000 patients die each year due to infection with agents that are resistant to multiple antibiotics (WHO, 2011). In the United States, the Centers for Disease Control and Prevention (CDC) reports that over 2 million people acquire serious resistant infections and at least 23,000 die as a direct result of these infections, with many more dying of conditions complicated by the antibiotic-resistant infection (CDC, 2013).

The pace of antibiotic discovery is not keeping up with the rapid evolution of resistance in microbes. Many of the major pharmaceutical companies that had driven the antibiotic discovery process for decades have eliminated or reduced their discovery programs. The United States Food and Drug Administration approval of new antibacterial drugs decreased by 56 percent between 1998 and 2002 (Spellberg, 2004). So, while the crisis escalates, the commitment to drug development is declining.

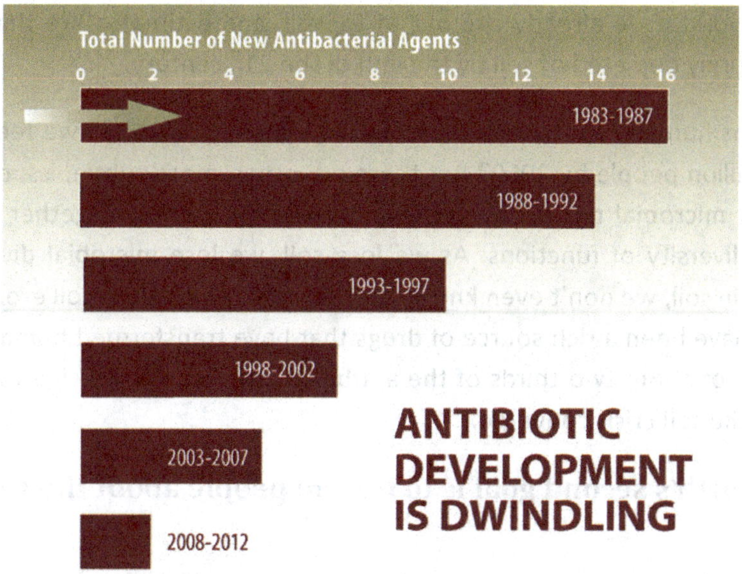

Source: *The Epidemic of Antibiotic-Resistant Infections*, CID 2008:46 (15 January)
Clin Infect Dis. (2011) May 52 (suppl 5): S397-S428. doi: 10.1093/cid/cir153

Tiny Earth's first goal is to educate students and the public about the antibiotic crisis.

Goal 2. The Soil Crisis

Most of us take for granted the Earth's crust that we call soil, but it is a precious and rapidly diminishing resource. Soil is well-known as the substrate in which we grow our crops, but it has shaped the course of human history by providing us with lifesaving drugs. In particular, most of the antibiotics we use to treat infectious disease come from bacteria in soil.

But despite its importance to our nutrition and health, we have squandered the Earth's soil. Disrupting it with plowing has degraded its structure, making it vulnerable to wind and water erosion. Many uses have depleted the soil's nutrients making it unhealthy and further vulnerable to damage by wind and water. When soil erodes, it moves from the place it was produced to the atmosphere or waterways. It can be transported to the other side of the Earth or into a river or other waterway. When soil releases its nutrients into water, the ecosystem becomes unbalanced and some organisms grow out of control. This has happened in the Gulf of Mexico where rich soil that eroded from farmland in the Midwestern United States is deposited after traveling down the Mississippi River. The nutrients that are carried along have caused massive blooms of algae that have depleted much of the Gulf of oxygen, producing a 7,900 square mile hypoxic (oxygen-free) zone where few organisms can grow—essentially a dead zone. So soil erosion is a serious problem both where the soil is lost and where it is deposited.

Soil forms slowly but erodes quickly. Most soil is produced at a rate less than 0.5 ton/acre/year, or less than one mm/acre/year. Some estimates suggest that it takes more than 100 years to produce one centimeter of soil. The frightening truth is that much of the world's soils erode at least 10 times this rate, and in some areas as much as 100 times as fast as the soil is produced. This is not sustainable.

Many areas of the world have already run out of topsoil, and estimates are that much of the United States is likely be entirely depleted of soil by the end of the 21st century.

Soil erosion threatens human existence. The obvious question is how will we feed a population that is predicted to top 9 billion people by 2050? But there are other, more subtle, issues that arise. Soil is the home to the richest microbial resource on Earth. Species are packed together at a high density and contain a fantastic diversity of functions. As we lose soil, we lose microbial diversity. Since we don't know all the species in soil, we don't even know what we are losing as the soil erodes. What we do know is that soil bacteria have been a rich source of drugs that have transformed human health. In particular, the soil is the source of about two-thirds of the antibiotics we use in medicine today. This is where the antibiotic crisis and the soil crisis converge.

Tiny Earth's second goal is to inform people about the soil crisis.

Goal 3. Original Research—the Essence of Science

The third goal of Tiny Earth is to engage students in original research. It is impossible to learn what science is and what scientists do without conducting research. Many students never experience science in its authentic form, but rather, read about what other scientists have discovered in textbooks. Knowledge discovered by others is important and scientists depend upon previous discoveries to guide and inspire their own research, but it is difficult for many people to get truly excited about the facts and concepts of science before they have experienced the thrill of research. There is nothing in the world like making a discovery. For a moment, you are the only person in the world that knows that tiny piece of information about the world around us. Cherish those moments. And then you get to share your finding, which has its own satisfaction. This is what being a scientist is about—having ideas, discovering new knowledge, sharing that knowledge, and putting in context with what is already known.

Tiny Earth's third goal is to engage students in original research.

Goal 4. Antibiotic Discovery from Soil

Although the antibiotics from soil microbes have been mined for decades, making them the premier source of antibiotics on the market today, they remain a rich treasure-trove of more drugs.

It may seem odd that the class of organisms intended as the target of antibacterial compounds is also a source of these lethal compounds, but it may also be that antibiotics serve functions other than as killer molecules in soil communities. Producing organisms may use antibiotics as signals to other species or they may provide specialized food sources. Whatever their roles in Nature, antibiotics have tremendous value to people for managing infectious disease, so we are fortunate that soil bacteria produced so many different compounds that we call antibiotics.

One of the challenges of identifying antibiotic-producing organisms is isolating the right ones in laboratory culture and then inducing them to produce their antibiotics. Much remains to be learned about the conditions that trigger antibiotic production, but we know that a small percentage of soil microbes produce inhibitory compounds under laboratory conditions quite readily—these are easy to find. One of your challenges will be to figure out new ways to induce other microbes to express normally silent pathways so that we can find the antibiotics that have remained hidden until now.

Can a team of student researchers solve the antibiotic crisis? The answer is a resounding, "yes!" To understand why we are so sure that **Tiny Earth's** students can discover new antibiotics, it's worth examining why antibiotic discovery has slowed to a trickle.

Pharmaceutical companies have said that the probability of identifying novel compounds is too low to be worth investment. The "low-hanging fruit," so to speak, has been picked. We disagree. There is copious evidence that there are many antibiotics waiting to be discovered, and students in our network have already identified novel compounds. We think it more likely that drugs prescribed for chronic conditions such as anxiety, depression, heart disease, and elevated cholesterol—drugs that people take for a lifetime—lured much of the pharmaceutical industry away from antibiotics that are prescribed for a relatively short duration.

While discovering and bringing an antibiotic to market is more than a semester-long process, we can certainly discover and characterize potential candidates in a few months—and learn about the research process along the way. By harnessing the power of undergraduate researchers, we can extend the discovery process to a wide range of geographic locations and a worldwide community of volunteers where every individual has the potential to make novel discoveries.

Tiny Earth's fourth goal is to discover new antibiotics.

Comments on "Tiny Earth"

"Tiny Earth" was chosen as the name of our network for several reasons. It captures both the small and the large—small microbes on a large planet, individual researchers part of large network, and many small discoveries that have a large cumulative impact. **Tiny Earth** also refers the Earth as a planet—which houses our network—and earth as soil, the precious resource that we study.

Perhaps the most important significance of **Tiny Earth's** name is that it alludes to how our network makes the Earth seem smaller because we knit together a global community of students and instructors with a common mission. Through our common database, we all work on the same dataset, created by our collective results, which could not be created by any one researcher or a small research team. This dataset will be informative, regardless of the outcome in terms of antibiotic production. Entries from around the world will help to identify trends about microbes that produce antibiotics and delineate geographic regions or biological habitats where antibiotic producers are abundant.

Science today is usually a collaborative effort by individuals in many laboratories. **Tiny Earth** takes that model to a new level, engaging tens of thousands of student researchers in antibiotic discovery and discovery about antibiotics.

References

Bulletin of the World Health Organization 2011;89:390–392 http://www.who.int/bulletin/volumes/89/5/11-088435/en/index.html

Antibiotic Resistance Threats in the United States, 2013. U.S. Department of Health and Human Services, Centers for Disease Control and Prevention (2013).

Spellberg, B., Powers, J. H., Brass, E. P., Miller, L. G., Edwards, J. E., Jr. (2004) Trends in antimicrobial drug development: implications for the future. Clin Infect Dis 38:1279–1286 doi: 10.1086/420937 pmid: 15127341.

How to Use the Research Guide

This research guide is designed to provide background information for each experiment in your research project. As a student researcher, you will be tasked with formulating your own hypotheses, designing your own experiments, picking which protocols to use, and interpreting and presenting your findings. As with every authentic research project, you will not have all the answers upfront. This guide is not meant to give you a set of instructions on how to execute your experiment or tell you everything you need to know about the subject. Instead, it will provide you with general **concepts**, **principles**, and **prompts** to guide you through the project.

Concepts

The **concepts** include a small selection of general biological and microbiological information, brief descriptions of what you would normally find in your textbook and gain from the lecture portion of your course. Each section, we will give you enough information to start thinking about the topic at hand and give you a basis to pursue more information. Use this opportunity to pursue information in a more targeted and goal-oriented manner. Ask yourself, what else would be relevant to know for this experiment or to answer the biological questions? By doing so, you will start to think like a scientist and take ownership of your research.

Principles

The **principles** are general microbiological guidelines to perform your experiments. We provide you with examples of microbiological techniques to show you some of the traditional approaches to answer the biological questions. The protocols section at the back of the book contains an assortment of traditional microbiological methods. After reading each section, you will be prompted to pick a protocol. In some cases, your instructor will provide a protocol or you will be asked to design your own or find one in the literature. While some experiments will have well-established protocols associated with them, you must always determine what you are trying to accomplish with your experiment and formulate your idea of "best results." Maybe you disagree with the traditional approach and can think of alternative ways to design the experiment and obtain even better results. We encourage this evaluation process because it will help you practice the kind of scientific thinking needed to execute any type of research. This is also how you will contribute to the collective creativity of the **Tiny Earth** network.

Prompts

Finally, each section in the research guide will feature several **prompts**. They will be scattered in the section or be presented as Biological Questions or Experiment prompts. At the end of every section, you will come across the topic or the goal of the experiment at the heading and the Biological Questions below it. This is where the *guiding* part of the research guide comes in. With every phenomenon we want to understand, or issue scientists wish to address, we come up with a list of questions. *Why is the* *sky blue? How will I solve world hunger? What is the matter with dung beetles?* We present some of the main questions for you, which should be evident after reading the background information in the section and browsing through some of your course materials. Use these questions to make informed hypotheses and design your experiments with specific outcomes in mind. The prompts will encompass the premise and the objective of the experiment, but you must tease those things apart yourself to start thinking like a scientist about own your project. Which takes us to the Experimental prompts. With every experiment, you will need to answer prompts, which are necessary to keep track of and record your research as well as validate your scientific approach and your findings. Should you need more room to write answers than what is provided (you probably will), use your own lab notebook. The following prompts are examples of the main anatomy of the Experimental Sheet at the end of every section:

- **Background:** What led to the biological questions at hand? What is the issue or the phenomenon you want to learn more about?
- **Objective:** List the objective(s) of the experiment. What do you wish to learn about the issues or phenomena at hand?
- **Hypothesis:** Come up with a temporary answer to your question and propose an outcome for your experiment. What do you propose is the explanation for the issue or the phenomenon at hand?
- **Experimental design:** Use text and graphics to explain your scientific approach. How will you go about testing your hypothesis?
- **Techniques/protocols used:** List the protocols you will refer to in the protocols section, or give a detailed description of alternative protocols (designed by you, provided by your instructor, or acquired from the literature). If you make any modification to the protocols provided, make sure you list them here.
- **Results:** Describe the findings of your experiment. What observations did you make? What data did you collect? Do not leave out your negative results—sometimes they are the most important part of your experimental findings that tell you which hypothesis is wrong or which new directions to pursue.
- **Interpretations**: What do your results mean? Speculate as to why you obtained the results you did, referring back to the original question and hypothesis. This will be the place to describe similar trends in your class results or the literature, and come up with insights about what your results tell you. This will also be the place to reevaluate your hypothesis and propose future directions.
- **Conclusion:** Wrap up your experiment with a brief summary and some of the new things you learned.

Use this resource to guide you through the research project and capture your sense of inquiry, observation, critical thinking, problem solving, and creativity. This is a guide into uncharted territory and only you will be the expert on your findings. Enjoy the ride!

Lab Safety and Best Practices

Knowledge and awareness of rules for personal and biological safety are highly important to creating a safe, organized, collaborative, and productive lab environment. Working in the laboratory requires constant surveillance for hazardous materials and practices. It also requires that we be able to identify potential hazards and prevent or manage them properly without compromising our safety and the safety of others. Lab etiquette informs us on how to behave properly in the lab by developing safe practices and habits, as well as effective responses to potentially hazardous activities and outcomes.

General Lab Safety and Etiquette

Entering and Leaving the Lab

1. Always wash your hands before and after doing lab work, and immediately after removing gloves. Never leave the lab without washing your hands first.
2. Never place personal belongings on benchtops or surfaces used for lab work. Keep clothing, backpacks, and other personal belongings in specified locations.
3. Food and drink are strictly prohibited in the lab!
4. Never use personal belongings, such as cameras, tablets, or utensils, for lab work. If use of personal cameras or tablets is required, ensure that they are completely sealed in bags or plastic wrap. When finished, dispose of bag or plastic wrap in designated waste containers before leaving the lab. Personal belongings that become contaminated with microbiological materials need to be properly sterilized (for example sprayed with a 70% ethanol solution) before leaving the lab.
5. Never remove lab materials, reagents, or cultures from the lab. Lab materials could introduce biologically and ecologically hazardous materials to the environment.
6. Know the equipment and supplies on your bench at all times. When in doubt about the identity of lab materials, ask your instructor.
7. For your own safety and the safety of others, know where eyewash and emergency shower stations, fire extinguishers, and emergency exits are located.

Personal Protection

1. Wear proper clothing. Closed-toed shoes and long pants are mandatory to enter the lab. Ensure that your hair is held in pace or tied back. Avoid clothes that expose too much skin or that are too loose. Clothes and hair can get in the way of flames, lab materials, and experiments.
2. Inform your instructor if you are immunocompromised, have allergies (e.g., latex allergy), or feel ill. Should you have any minor cuts or non-intact skin, ensure that they are covered with waterproof bandages. When in doubt, inform your instructor.
3. When carrying out experiments and handling lab materials, it is mandatory to wear personal protective equipment (PPE):

a. gloves
b. protective eyewear (goggles or spectacles)
c. lab coat

4. Always check gloves for integrity. Gloves provide only a thin layer of protection and may contain holes or pores that could allow the passage of fluids and solid materials.
5. Avoid touching shared surfaces, such as light switches, door knobs, and handles, while wearing gloves.
6. When removing PPE, avoid touching the outside of the gloves or your lab coat. Dispose of used gloves in designated waste containers, not regular trash, and immediately wash your hands. Remove eyewear with clean hands.

Handling Lab Materials and General Precautions

1. Handle Bunsen burners or flames with caution: before lighting flame, remove flammable objects (paper, plastic, solvents) from its range.
2. When working with chemicals, such as organic solvents, work in a fume hood and never near a flame.
3. Avoid bringing samples, cultures, reagents, and hazardous equipment close to the face or skin. Never ingest, inhale, lick, or taste materials in the lab!
4. Practice proper pipetting and never mouth-pipette. Never point the pipette toward your body or others. Dispense liquids down the sides of tubes/containers and as close to the content as possible to minimize splashing.
5. Keep test tubes and microcentrifuge tubes in racks at all times.
6. Minimize aerosols by handling liquid cultures, reagents, and solutions carefully. Avoid excessive shaking and forming bubbles in uncapped tubes when pipetting, pouring, decanting, mixing, etc.
7. Cap liquids when not in use, and always cap when mixing, vortexing, or inverting. This measure will help prevent cross-contamination and avoid you coming into contact with potentially hazardous biological materials, fluids, and aerosols.
8. Cap or cover samples, cultures, and specimens when not in use, and keep in specified locations. Avoid close contact with face, skin, and clothes. Avoid stacking plates and leaving them in the middle of the benchtop when not in use.
9. Label samples and cultures with your name, date, and content (See notes on labeling in next section). Make sure that all samples, cultures, media, reagents, solutions, and used test tubes, microcentrifuge tubes, and conical tubes, have proper labeling. Avoid using unlabeled materials.
10. Disinfect surfaces before and after every lab activity. Wipe benchtops and large exposed surfaces with disinfectant solution. Apply 70–85% ethanol or 1–10% chlorine bleach solution, or any disinfectants provided by instructor. Never apply disinfectant solution via spray or squirt bottles near an open flame!

Disposal and Spills

1. Place used or contaminated lab supplies in proper waste containers. Locate the regular trash, biological waste containers, sharps, and broken-glass containers. Any lab supplies that come

into contact with cultures, media, reagents, or any potentially hazardous material should not be left on the benchtop.

   a. Place used reusable equipment in proper receptacles for autoclaving; this includes metal loops, hockey sticks, glass beads, forceps, and glassware.
   b. Dispose of samples, cultures, and media in designated biological waste receptacles.
   c. Dispose of used weigh boats, Petri dishes, and other disposables in designated biological waste receptacles.
   d. Dispose of used sharps (slides, coverslips, Pasteur pipettes, broken glass, toothpicks, wooden sticks, plastic tips, etc.) in designated sharps and/or broken-glass containers. Ask your instructor about glass (e.g., microscope slide) and/or broken-glass disposal.
   e. Ask instructor where to dispose of large pointy objects, such as serological pipette tips.

2. Properly dispose of contaminated media and plates.
   a. Do not remove the lid from plates with fungal growth. Plates with yeasts or molds disseminate spores, which can rapidly spread and contaminate the lab. Consult your instructor for inspection and proper disposal.
   b. Do not use agar plates (solid media) or liquid media that are supposed to be sterile but appear to have microbial growth. Consult your instructor for inspection and proper disposal.
3. Immediately clean spills by flooding the area with disinfectant for a few minutes and wiping with appropriate absorbent material. For large spills, immediately inform your instructor.

Behavior

1. Always be respectful of your lab mates and their workspaces.
2. Do not distract or obstruct lab mates when they are carrying out an experiment as this can compromise their safety and the integrity of their experiments.
3. Communicate questions or doubts about lab safety and proper technique at all times. Also, say something when others are not adhering to lab safety rules and nonhazardous techniques.
4. Always keep written records of your experiments. If you are lost or confused, do not hesitate to ask.
5. Be aware of yourself and your surroundings at all times.

American Society for Microbiology Guidelines for Biosafety

When working with environmental samples, we are always faced with unknown microorganisms. Based on what we know of soil bacteria, most will probably be harmless or nonpathogenic. These bacteria are treated as low-risk or biosafety level 1 (BSL1) microorganisms. However, bacteria that are normally nonpathogenic may become pathogenic under unusual conditions or environments. Furthermore, we can never disregard the possibility of isolating a pathogen. These bacteria are treated as moderate to high risk or biosafety levels 2 or 3 (BSL2/3), respectively. **For these reasons, we should always treat bacterial isolates in the lab as potential pathogens.**

Another risk of isolating unknowns is increased bacterial load. Every time we plate a soil sample, we are allowing individual bacterial cells to reproduce exponentially under optimal conditions. Colonies contain millions of cells, where only a small number may be necessary for infection. Students with a greater risk of infections – due to compromised immune systems or prolonged sickness – should inform their instructor.

The following guidelines, based on the ASM *Guidelines for Biosafety in Teaching Laboratories* (2012) contain precautionary measures that we must take when working with unknown isolates:

1. Treat all unknown bacterial isolates regardless of sample source (nature, household items, human skin) as potential pathogens or microorganisms that may need BSL2 containment, and, in rare cases, BSL3 containment.
2. BSL1 labs may plate bacteria from environmental samples; however, these plates must remain covered and sealed or stored in a secure location. Isolates must remain enclosed at all times, and only observed through the lid. After observation, plates must be decontaminated by autoclaving and properly discarded.
3. BSL2 labs may subculture, or transfer, unknown bacterial isolates from the original medium (described above) to a fresh medium for further testing.

Adherence to the aforementioned guidelines, along with other best-practice suggestions, aseptic technique, and general laboratory safety and etiquette, will help promote a safe and productive lab environment.

For more information:

- **Please read through the ASM *Appendix to the Guidelines for Biosafety in Teaching Laboratories*:** http://www.asm.org/images/Education/FINAL_Biosafety_Guidelines_Appendix_Only.pdf
- **Please visit the following URL for more information on guidelines for biosafety teaching:** http://www.asm.org/index.php/education2/22-education/8308-new-version-available-for-comment-guidelines-for-best-biosafety-practices-in-teaching-laboratories
- Emmert, E. Biosafety guidelines for teaching laboratories. 2013. JMBE. 82, 14. Available at http://jmbe.asm.org/index.php/jmbe/article/view/531

Safety Rules

1. Treat all environmental samples as a potential source of pathogenic microorganisms.
2. Whenever culturing from soil, the use of selective media with an antifungal agent (50 µg/ml cycloheximide) is recommended.
3. Never culture environmentally isolated and unidentified microorganisms at 35 °C or higher to reduce (but not eliminate) the chance of culturing human pathogens, many of which require 37°C for growth.
4. Never culture environmentally isolated and unidentified microorganisms in broth (liquid culture). All experiments with unknown microorganisms must be done on agar.
5. Properly dispose of plates with fast-growing microorganisms, which spread from a single colony to cover most of the surface of the plate.
6. Keep plates covered at all times. Plates can only be observed through the lid.
7. When handling bacteria:
    - Follow "Aseptic Technique" section.
    - Wear appropriate PPE and follow the lab safety rules.
    - Keep a reasonable distance between your body and the plate.
    - Keep the lid at an angle over the plate to insert a pipet tip, toothpick, inoculating loop, or any other sterile inoculating supply.
    - Dispose of or place any used or contaminated supplies in designated BSL1/2 waste container or ethanol/bleach solution.
    - If your gloves touch any bacterial growth or become compromised, quickly cover the plate, stop what you are doing, dispose of your gloves, and thoroughly wash your hands with disinfectant soap.
8. Do colony PCR and sequencing of unknown antibiotic-producing microorganisms before performing biochemical tests or chemical extractions.
9. Clean the workspace with ethanol or bleach solution at the end of lab session.
10. When in doubt, do not do anything you feel is unsafe and contact your instructor or lab safety officer.

Aseptic Technique

Aseptic or sterile technique is a central concept in microbiology and may be the most important part of working in a microbiology lab. The goal of aseptic technique is to promote practices that (1) prevent the contamination of cultures, lab supplies, and equipment with bacteria or fungi from the environment, and (2) prevent the contamination of individuals working in the lab with potentially pathogenic bacteria. Aseptic technique requires constant attention until it becomes second nature. Practice and strict adherence to these practices is crucial to making proper aseptic technique a habit; the health of individuals in the lab and the integrity of experiments depend on these principles. Here are some general guidelines for practicing aseptic technique:

- Remember that bacteria are ubiquitous in the environment and our bodies.
- When possible, light a Bunsen burner or flame when working with both solid and liquid bacterial cultures. The Bunsen burner will create an umbrella of sterility by creating an upward airflow in your workspace and preventing debris from falling on your sterile materials. Work at a safe distance from your flame without risking burning yourself or anything in your workspace.
- Handle lab materials sensibly, avoiding unnecessary motions or contact with contaminated objects/surfaces or things that are not pertinent to the protocol at hand. Inoculating loops, sticks, toothpicks, pipette tips, serological pipettes, agar plates, media, and other previously sterilized materials are packaged and stored under aseptic conditions. Should you inadvertently touch any items used in the protocol, the item is no longer sterile and must be resterilized or discarded (if the item is disposable).
- Keep in mind that Petri dish lids prevent particulate matter and airborne bacteria from contacting the plate surface, while allowing the diffusion of air around the edges. Whenever a plate lid is removed, it should be held over the plate as a shield. Do not place the lid face down on the benchtop; avoid placing it on the bench all together. Do not leave plates uncovered. Do not walk around the room with an open plate.
- When working with cultures in test tubes, work diligently and as fast as possible. If Bunsen burners are available, pass the uncapped mouth of test tubes through the flame to prevent air from coming into the test tube. Keep tubes open for a minimum amount of time. While lids are removed, hold the tube at a slight angle so that airborne contaminants cannot fall into the open tube. Point tubes away from your face and body while working with them.

Recordkeeping

An important aspect of scientific research is effective communication of one's findings. Today's scientific problems are not solved in isolation but rather by groups or communities of researchers working together or building on the results of others. Methods of communication include publication in peer-reviewed journals, presentation of results at public forums through posters or seminars, submission of data to databases such as GenBank, and communication to the public through media outlets or informal gatherings. In order for research to be accepted by other scientists it must be reproducible. Good recordkeeping is imperative for ensuring that others can replicate one's experimental procedure. **Tiny Earth** is a real research project with a worldwide community of undergraduate researchers. To this end, it is important to keep accurate records. This includes the upload of required information to the database, documenting experimental procedures in your lab notebook, and recording digital photographs of results.

Tips for keeping a laboratory notebook

- Notebook entries should be made in ink and in chronological order. Entries should not be erased or whited out. If an entry contains an error, a line should be drawn through the error and new text should continue in the next available space.
- Start each entry with a title and a clear description of the objective of the experiment.
- Date each experiment. The dates when the work was begun and completed should be recorded.
- Record all materials used in the experiment.
- Record experimental design as well as protocols used. Any modifications to existing protocols should be thoroughly explained. Never assume that any protocol is common knowledge. Be sure to write your own descriptions and definitions.
- Record all data collected and observations made in the experiment. If possible, include photos and diagrams.
- At the end of the record, evaluate your experimental results in an interpretation or discussion section. Draw conclusions and future directions, if possible.
- Explain all abbreviations and terms that are nonstandard. Explain in context, in a table of abbreviations or a glossary.
- Document everything; take detailed notes and photographs during lab. Immediately download and annotate photos to capture accurate data.
- Record negative results. Scientists, especially novice scientists, can learn a great deal from negative results. Reviewing the protocol, brainstorming where the error occurred or reasons for the negative results are key teaching and learning moments. Take advantage of them.
- In general, a good practice is to organize your report as soon as you obtain data. Summarize protocols, experimental designs, and results, make presentable tables, and incorporate the photos taken during lab.

Labeling

Labeling materials and supplies is a critically important aspect of science and lab safety. With multiple plates, tubes, and samples, it is easy to become confused and lose track of things between and within experiments without the proper labeling practices. Here are a few guidelines for labeling:

- Always label the back of the Petri dish, never the lid, because the lid may become separated from the actual dish containing the growth medium (for example, if plates are dropped).
- Always add a date. Dates help orient the scientist as to when the experiment was performed and provide information about the quality of materials.
- Write as legibly as possible – others may need to find your materials or experiments.
- Label experiments with your initials and experiment number and number of trials, for example: TMT 2.1 – experiment 2 performed the first time.
- Label isolates with the your initials, type of medium used, and an identifying number. For example: SS-PDA-5 stands for Sam Smith, potato dextrose agar medium, isolate number 5. If the identity of the organism is known, label with scientific nomenclature: genus, species, and strain.

Entering Data into the Tiny Earth Database (TED)

The power of **Tiny Earth's** student research is substantially greater than the sum of its parts. When student research data are recorded in TED, **Tiny Earth's** Database, our network's discovery potential expands well beyond new antibiotics. Given **Tiny Earth's** global footprint and the diversity of soils mined by **Tiny Earth** student-scientists, soil collection and soil isolate data entered into TED will help us answer questions and better understand factors influencing antibiotic production in soil. With TED's help, **Tiny Earth's** potential for *studentsourcing antibiotic discovery* swells.

Section 1: Living on a Bacterial Planet

Microbes and the Environment

Microbes live in diverse natural habitats, ranging from hot hydrothermal vents deep in the ocean to the insides of our guts and the surfaces of our computer keyboards. The ubiquity and versatility of microbes is the result of their long, intimate relationship with our planet, dating back nearly four billion years. Throughout this long journey, microbes have been active contributors—perhaps the most important drivers—in the transformation of our planet and evolution of its inhabitants. They have shaped the biosphere, driving the evolution of complex cells and multicellular organisms that continue to be largely dependent on the specialized products made by bacteria. Microbes have influenced the Earth's biogeochemical systems, driving the cycling of elements that would otherwise be inaccessible to other systems and organisms. Elements such as carbon and nitrogen would only exist in forms we cannot assimilate into our bodies without them. Microbes living more than 2 billion years ago produced the oxygen that made the Earth's atmosphere transition from anaerobic (oxygen free) to aerobic (oxygen rich), and created the ozone layer that protects the Earth from UV radiation. Microbes might even play a role in forming water droplets that produce rain in clouds.

Definitions

Microbial ecology is the study of microbes' interactions with each other and with their environments. Microbes are part of complex ecosystems, and the communities within those ecosystems teem with life and bustle with activity. Microbes communicate, exchanging and sharing instructions to make new compounds. They may compete for a limited resource or cooperate to maximize the availability of another. Two major focuses of microbial ecology relate to the **biodiversity** and **bioactivity** in an ecosystem. These terms refer to the types of organisms present in an ecosystem and what their activities entail, respectively.

Microbe is a general term for an organism that cannot be seen with the naked eye, and is therefore microscopic. Microbes include **bacteria** and **archaea (prokaryotes)**, and microscopic **eukaryotes**, such as yeasts, molds, and protists. **Viruses** and **prions** are also lumped into the broad category of microbes. However, these agents do not meet the all the basic requirements for life, such as the ability to reproduce on their own, and are therefore considered nonliving. Prokaryotes and eukaryotes differ in their level of complexity and cellular organization. Eukaryotes package their genetic material inside a compartment called the **nucleus,** whereas prokaryotes do not. Prokaryotes do not contain intracellular compartments or organelles and maintain their genetic material in the cytoplasm.

Living on a Bacterial Planet

The number of individual prokaryotic cells in soil across the planet has been estimated to be 26×10^{28} (260 billion billion billion) (Whitman, Coleman, & Wiebe, 1998). The collective impact of prokaryotes, such as bacteria, on our planet cannot be overstated.This course will focus primarily on bacteria because:

- *Bacteria are virtually everywhere on the planet and are highly abundant – in many ways we live on a bacterial planet;*
- *Bacteria are great research tools because, in comparison with other organisms, they are simple but make specialized chemical compounds, many of which have human applications;*
- *And bacteria greatly impact our health – some may cause infectious disease and others protect our bodies from invasion by harmful bacteria.*

Getting Started

In this course, we will conduct research involving soil bacteria and ask questions pertaining to microbial ecology such as:

- *What types of bacteria are present in our soil samples?*
- *How can we accurately count bacteria and tell them apart from one another?*
- *How many different types and/or individual bacteria of a given type are there?*
- *How do environmental factors influence these numbers?*
- *Does isolation in culture influence bacterial activity?*
- *Are the activities of a particular bacterium beneficial or harmful to another bacterium?*
- *Do they produce specialized molecules like pigments, toxins, or antibiotics?*
- *How active are their antibiotics and what do they target?*
- *Are they resistant to the effects of other antibiotics?*
- *Can we manipulate their production of antibiotics?*
- *Are these antibiotics useful for humans?*
- *Are they known or novel antibiotics?*

We will design and execute experiments to answer these questions, test hypotheses, and develop new insight into the microbial world. As microbiologists, we will refer to many disciplines in the natural sciences, ranging from cell biology to biochemistry to analytical chemistry to genetics. Tools developed in these disciplines will be very important for us over the following weeks as we assess the biodiversity and bioactivity in our soil samples. We will be challenged to think critically about microbial interactions and functions, as well as about ways that we can apply our newly acquired knowledge to enrich the existing body of knowledge and improve human welfare.

References

Whitman, W. B., Coleman, D. C., & Wiebe, W. J. (1998) Prokaryotes: the unseen majority. Proc Natl Acad Sci U S A 95(12):6578–6583.

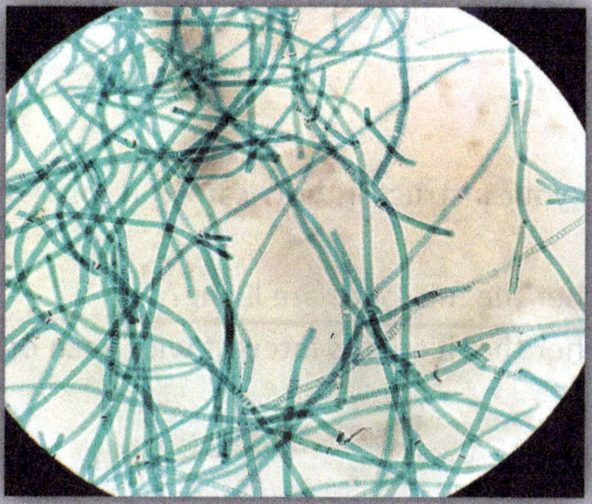

Living on a Bacterial Planet

This photo, taken through a light microscope eyepiece, shows the unique shape and color of the **photosynthetic cyanobacteria** (also known as blue-green algae). Cells form long chains (filaments) and inhabit a variety of habitats, including soil and oceans. Cyanobacteria produce oxygen through photosynthesis and also chemically modify gaseous nitrogen from the atmosphere, which they use for growth. The evolution of cyanobacteria 2.5 billion year ago was perhaps the most transformative event in earth's evolutionary history. Massive amounts of oxygen were released into our planet's atmosphere for the first time, drastically changing its composition. This event triggered the mass extinction of many oxygen-sensitive microorganisms, enabled evolution of oxygen-dependent respiration, and gave rise to the ozone layer, which allowed life to migrate to land and evolve into the wide biodiversity (the human species included) we see on earth today.

Photo source: commons.wikimedia.org

Experiment 1: Devise a method to transfer microbes from a soil sample to a medium in the lab

Biological Questions:

1. Given a soil sample from your instructor, how can you isolate microbes from soil into a medium in the lab?
2. How will you know whether the microbes are living?
3. How will you know whether the microbes interact with each other?

Background:

Objective of the experiment:

Hypothesis and rationale:

Isolation method used:

Observations:

Evaluation and conclusions:

Notes:

Section 2: More Than Just "Dirt"

Soil

Throughout human history, humans have looked at soil with a sense of mysticism and spirituality, in appreciation for its necessity for life. Soil sustains humans through agriculture and provides a platform for our daily activities. In many respects, soil not only sustains us, but also makes us who we are. The development of civilizations has depended upon soil health; poor soils have resulted in the collapse of entire civilizations, whereas fertile soils have enabled civilizations to flourish. As a symbol of fertility and the raw materials of life, soil has inspired many of the world's religions and traditions.

Soil is an extensive natural body covering most of the Earth's terrestrial surface and mediating the Earth's systems in the biosphere, atmosphere, hydrosphere, and lithosphere. Most evident of all, soil supports forests and plant life, which in turn produces half of the oxygen in the atmosphere and 95 percent of the world's food. Soil greatly influences the planet's ecosystems but is itself affected by natural forces and human, or anthropogenic, activities as well. Humans have rapidly extracted large quantities of soil constituents that were accumulated over millions of years through soil genesis and mineral weathering and sedimentation. Human activity can disrupt the natural cycling of elements and can pollute the soil with extraneous chemicals that natural ecosystems are not well-suited to remove or not well-suited to remove or degrade. All these external factors greatly influence the life that depends on soil and can produce predictable changes in the biodiversity, bioactivity, and biomass we observe. This is because soil is itself an ecosystem for some of the most numerous, diverse, and dynamic organisms on Earth.

Soil Characteristics

The main characteristics defining soil are its physical structure, chemical composition, and association with plant roots or other resident organisms. The association of soil with plant roots creates a distinction between soil types that is often referenced by microbiologists and agronomists. We can distinguish between the **rhizosphere**, or soil that is in direct contact with living plant roots and their secretions, and **bulk soil**, which constitutes everything else. The microbial profile of the rhizosphere differs starkly from that of bulk soil. Root systems create microenvironments that favor the growth of specific bacteria. These microenvironments affect the way nutrients and energy are delivered, how water drains, and how temperature fluctuates. Different plants and microbes have specific chemical and metabolic signatures, which in turn change the chemical properties of soil. This interplay between living and nonliving entities defines the characteristics of their respective environments.

The physical structure of soil is determined by the proportions of three main particle types: sand, silt, and clay. These particles clump together to form aggregates of different sizes and textures. Soil with large, coarse particles (sand) or clumps is more permeable than finer soil, allowing the passage of water and faster leaching of soluble molecules. This can deplete the soil of necessary nutrients and limit the

Tiny Earth
studentsourcing antibiotic discovery

amount of life that can be sustained by it. Conversely, soils containing more fine particles (silt and clay) or smaller clumps are less aerated than coarse soils, restricting the growth of microbes that need oxygen gas (O_2) below the surface.

The rates at which materials are deposited into the soil and subsequently removed or processed by natural forces, such as living organisms and water, drastically impacts the spatial composition of soil. As we dig into the soil, getting further away from the surface, we encounter differences in physical and chemical composition, as well as differences in the microbial populations found at each depth. Cross-sections of most soils reveal that they are divided into distinct layers called **horizons** (Figure 2-1). Understanding the different horizons is important to categorizing soils. The structure of horizons is also a good predictor of the nutritional characteristics of the microbes living within them. Generally, as we descend from one horizon to the next, the abundance of organic components, as well as the exposure to light and oxygen, declines.

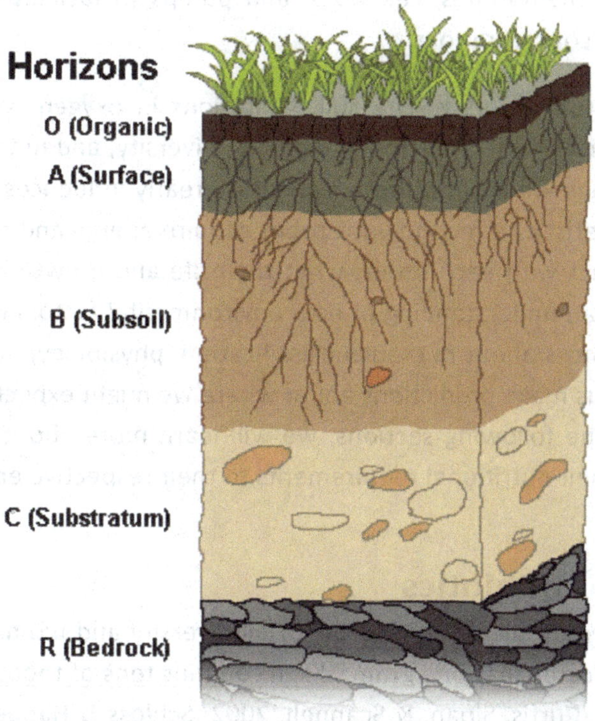

Figure 2-1. Soil Horizons: Soil is divided into layers called horizons, which vary in their physical and chemical characteristics. As we descend from one horizon to the other, organic matter and microbial life decrease and we approach greater accumulations of inorganic materials such as minerals and rock. Source: https://en.wikipedia.org/wiki/Soil_horizon

The **O Horizon** sits at the surface of the soil profile, where un-decomposed plant matter is directly exposed to the elements. This is what we typically see on the forest floor. Right below this thin layer sits the **A Horizon**, or **topsoil**, which is rich in organic matter and contains most soil life. Microbes actively decompose organic material to form humus, and intimately interact with plant roots and other organisms such as fungi, insects, and worms. Some of the organic and inorganic components of this layer are easily drained by water and accumulate in the subsequent **B Horizon**, or subsoil. The B Horizon contains lower levels of microbial life and a greater proportion of anaerobic bacteria that do not require oxygen for respiration. Inorganic matter such as minerals and partially weathered parent material constitute the **C Horizon**. In this layer, microbial life is low in abundance but nevertheless present. Researchers have even found microbes in bedrock, thriving in an environment free of sunlight, oxygen, organic molecules, and with very little available water.

pH is a significant factor in shaping microbial life in soil. Plants and microbes can only survive within a specific pH range. This means that pH can be viewed as reflective of soil health. After comparing different environments across the globe, Fierer and Jackson found that soil pH was the best predictor of microbial diversity. Arid soils, which are poor in water and nutrients but generally have a near-neutral pH, were richer in microbial diversity than rainforest soils, which are usually more acidic (2006). This is in spite of the fact that rainforests comprise incredible amounts of diversity at the **macroscale**. Another important factor is the ionic concentration of the soil. Microbes must be equipped to withstand **osmotic forces** with well-adapted membranes, cell walls, and pumps to facilitate the movement of ions and resist fluctuations in ionic concentrations.

As we approach the bottom layers of the soil, limitations in oxygen, sunlight, carbon sources, and changes in pressure drastically affect microbial biomass, diversity, and nutrition. It is important to know which layer of soil microbes are isolated from since this greatly influences microbial growth conditions. Furthermore, soil characteristics affect how microbes acquire energy and nutrients, how they adhere to surfaces, how they interact with each other, what their life and growth cycles are like, and how they evolve over time. Having an understanding of how environmental factors influence microorganisms can help when making inferences about microbial classification, physiology, and basic cellular biology. This knowledge can also help us make predictions about where we might expect to find the most biodiversity and active microbes. In the following sections, we will learn more about microbial nutrition and how microbes have adapted their nutritional requirements to their respective environments.

Soil Microbes and Antibiotics

Soil harbors an abundance of microbes in terms of biodiversity and biomass (the number of individual organisms). Research indicates that <u>one gram of soil contains tens of thousands of bacterial species and billions of individual cells</u> (Curtis, Sloan, & Scannell, 2002; Schloss & Handelsman, 2006). **Soil microbes**, or microbes that live in the soil, have evolved many adaptations to their environment. They synthesize an arsenal of biochemicals used for bacterial warfare, signaling, or acquiring nutrients. They break down complex molecules, such as those present in decaying plant matter or carcasses, and cycle nutrients and elements back into their ecosystems. In addition, many soil processes that decompose biological materials release greenhouse gases and characteristic odors. It is no wonder that humans have ascribed so many mystical qualities to soil.

Soil bacteria live in a crowded and highly competitive environment where conditions constantly change and resources are limited. In order to cope with a changing and competitive environment, many microbes have evolved specialized molecules that mediate their interactions with their surroundings. Pigments, toxins, and antibiotics are specialized molecules known as secondary metabolites. **Secondary metabolites** are organic compounds produced by bacteria that enhance their chances for survival, while remaining nonessential for fundamental growth processes (unlike structural molecules and DNA, for example). Antibiotics are secondary metabolites that inhibit the growth of other microorganisms. In their natural habitat, some microbes may produce antibiotics to reduce competition for resources; these same antibiotics save human lives when used as drugs to treat infectious disease (more on secondary metabolites and antibiotics in Section 9).

The majority of antibiotics in commercial and clinical use today are derived from soil bacteria. Actinomycetes, which are common soil bacteria, produce 60 percent of clinically important antibiotics. There is still tremendous potential for finding new antibiotics among the immense soil reservoir of microbes and molecules; in fact, the more scientists study soil, the more we realize just how little we know about the chemistry of soil microbes. The vast unknown features and intricacy of the soil makes this ecosystem a fascinating place to study microbes, which is the focus of our research project. We will collect soil and bring it to the lab to explore the capabilities of the microbes living in it. Keep in mind that a soil sample you collect represents the most complex ecosystem on Earth. Your soil sample may not encapsulate the full range of biology and chemistry of soil, or even of the soil immediately adjacent to the sample you removed, but it will be very, very complex. So pay close attention to the characteristics of the soil ecosystem and what they might indicate about the microbes inhabiting it.

- *How will we know whether our sampling location is rich in abundance and diversity of antibiotic-producing bacteria?*

As a collaborative network of student researchers and instructors across the globe--from Botswana to Connecticut, U.S.A, **Tiny Earth** is poised to collect data at a scale never before reported. By documenting detailed information about your sample location and keeping records of your observations (the number of microbes isolated, the number of those that produce antibiotics, etc.) we will begin to amass a large and useful data set that may reveal trends in soil. This information may prove helpful, not only for future classes, but also for the research community as a whole, to distinguish likely sources of new antibiotics from those that have a low probability of yielding antibiotic producers.

Experiment 2: Find a local soil environment you wish to sample

Biological Questions:

1. What soil environments will give us the most unique and greatest diversity of bacteria to study in the lab? Define what unique and diverse means to you.
2. What local soil environment will you choose for sampling and why?
3. What do you hypothesize about the abundance and diversity of microbes in your soil sample and why? How will you test this? Think of controls to use in your experiment.

Background:

Objective:

Criteria for picking soil sample:

Soil sample site:

Hypothesis and rationale:

Technique used for collecting your soil sample:

Notes:

Soil Sample Data Collection Sheet

Worksheet also available in protocols section. Your instructor may only want you to record data on some of these or may indicate additional data they want you to record.

Collected By:	
Date of Collection:	
Depth:	
Type of Soil:	
Temperature of Air:	
Temperature of Soil:	
Weather Conditions on Day of Collection:	
General Location:	
GPS Coordinates via Google Earth:	
Sample Site Descriptors:	
Additional Data to be Collected in Lab:	
pH of Soil:	
Water Content:	
Organic Content:	

Additional Documentation:

- Photo of sampling site
- Identification of any plant species observed near site
- Photo of soil in collection tube

Soil Sampling Kit Contents:

- Data sheets
- 50-mL conical tubes
- Lab marker
- "Scientific Soil Sample Collection Device" (plastic knife or spoon)
- Alcohol pads (used to disinfect collection device, thermometer, ruler before taking each sample)
- Thermometer
- Site-marking flag (if return to site necessary)
- Plastic ruler

Printout of "Guide of Texture by Feel": http://soils.usda.gov/education/resources/lessons/texture/

Section 3: Redefining "Growth" and "Culture"

Bacterial Growth

Bacterial cells "grow," in the sense of getting larger, to a limited extent (generally doubling in size before dividing), but this is not what microbiologists usually refer to when using the term "growth." We use "**growth**" to describe an increase in the size of a microbial population. In other words, when talking about single-celled organisms, "growth" usually means "reproduction." Reproduction is critical to the survival of any species, and bacteria are true survivors, reproducing exponentially under optimal growth conditions.

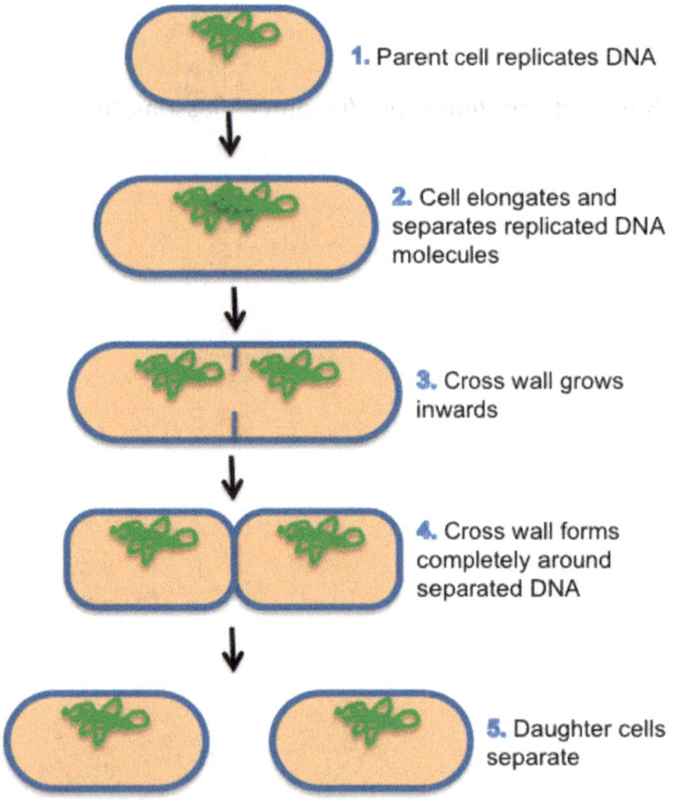

Figure 3-1. Schematic of binary fission, the process of bacterial cell division whereby a parental cell gives rise to two daughter cells. Every successive round of cell division doubles the size of the bacterial population. This results in exponential growth under optimal conditions.

The bacterial growth cycle is driven by complex intracellular and extracellular signals, which tell the cell when it is time to replicate. The first step in the process is DNA synthesis, in which a bacterium makes an extra copy of its chromosome, which contains all the genetic information necessary to give a cell its instructions for life. The cell then increases in size and elongates, pulling two identical chromosomes apart. Once the chromosomes are localized at the opposite ends of the elongated cell, the plasma membrane and cell wall start forming an "equatorial septum," which splits the parental cell into two daughter cells. This type of symmetrical cell division is known as **binary fission** (Figure 3-1). A few bacteria reproduce by an alternative mechanism known as **budding**, in which a small daughter cell "buds off" a larger mother cell. In either case, daughter cells remain genetically identical to the parent.

In any given population of bacteria, some cells will be dividing, others may be between stages or quiescent, and others may be dying. Yet when conditions are optimal for growth, such as a warm environment with just all the right nutrients, most cells will be dividing rapidly. The time it takes for a population of bacteria to double in size is called the **doubling, or generation, time**. In rich nutrient environments, *Escherichia coli* populations may double every 20 minutes, whereas other species of bacteria, such as *Mycobacterium tuberculosis*, may take up to 16 hours. "Optimal growth" conditions, however, are virtually nonexistent in natural habitats, where bacteria are kept in check by competing microbes and limited resources. Optimal growth conditions are typically created in the lab, where microbiologists have developed nutrient-rich environments to support the growth of some bacteria.

Bacterial Cultivation

To promote the growth of bacteria in the lab, which is referred to as **culturing** or **cultivating**, we need a medium to provide them nutrition and a place to live and reproduce. **Media** (plural of the word medium) are "bacteriological foods," which are mixtures of nutrients contained in liquid or solid form (more on media and nutrition in Section 4). When a microorganism is introduced into a liquid medium or the surface of a solid medium and allowed to reproduce, we refer to the resulting mixtures as **cultures**. Liquid media are typically referred to as broths, and solid media as agar plates (agar is the solidifying agent that is used to make the media solid in a Petri dish) (Figure 3-2) (more in Section 4). Culture conditions are carefully controlled parameters, such as temperature, humidity, and oxygen levels. However, there is no *one* food for bacteria. Just as bacteria are taxonomically diverse, so are their nutritional requirements. Every discovery of a new organism requires a careful assessment of the nutrients and conditions necessary to support its growth.

The greatest limiting factor in studying microbes is the small number of bacterial species that grow in culture. Until the 1990s, most of our knowledge of bacteria was based on the small number of so-called **culturable** bacteria, or bacteria that can be cultured in the lab based on current knowledge. This **culture-dependent** approach has been the basis for studying bacteria in most traditional microbiology labs. However, microbiologists have developed alternative approaches to get a wider and more accurate representation of bacteria living in the environment. These **culture-independent** approaches do not require bacteria to reproduce in the lab or exist in large numbers. It relies on advanced and expensive equipment to detect and analyze millions of microscopic entities individually. **Metagenomics**, the study of all the genetic material present in a given environment, allows us to sample soil and analyze all the DNA present in it as a signature for all the organisms living in it, whether or not they grow in the lab. These methods indicate that only 0.3 percent of all bacteria in soil can be cultured in the lab (Amann, Ludwig, & Schleifer, 1995). This means that only 3 in 1,000 bacteria can be studied in the laboratory by using a culture-dependent approach.

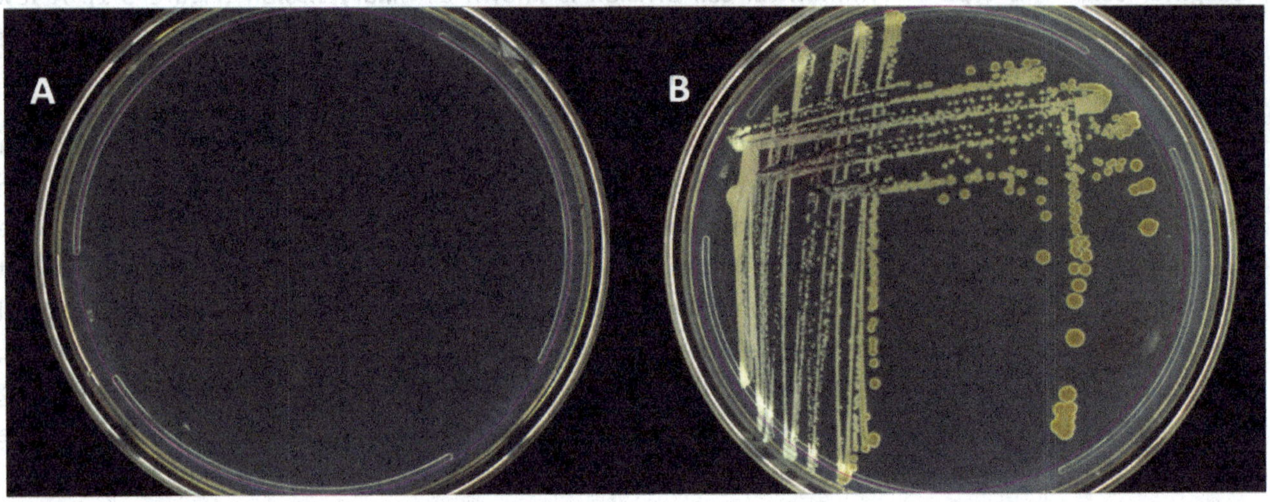

Figure 3-2. (A) A solid medium, composed of a mixture of nutrients, water, and agar (a solidifying agent) with no bacterial growth. (B) The same solid medium with a culture of *Staphylococcus cohnii* after 48 hours of incubation at 28°C.

The limitations of the culture-dependent approach result from inadequate knowledge of conditions that favor bacterial growth in natural habitats. Studying the soil metagenome in many ways has led to more questions than answers. Although the metagenome represents the complexity of the microbial world, its functional diversity remains difficult to study without culturing bacteria. In our lab, we will rely on the culture-dependent approach to study soil bacteria because it gives us functional information, can lead directly to antibiotic discovery, and has not been used to fully mine the soil or the antibiotic producing bacteria that inhabit it. Just remember that we won't be visualizing the entire microbial world with culture methods.

CFUs and Bacterial Colonies

The most important feature of the culture-dependent approach is that it allows us to study microscopic entities at the macroscopic scale. However, it is important to keep in mind the requirements and limitations of this approach:

1. *Only **viable** (alive and capable of reproducing) cells will grow. Dead and/or dormant cells will not be detected.*
2. *Only cells that are **culturable** (capable of growth on the selected culture medium and under the prescribed incubation conditions) will be detected. Cells that cannot grow under the selected laboratory conditions will not be detected.*

In order to detect a microorganism in the lab, we need a visual cue that will tell us whether the microorganism meets the above requirements. We are familiar with the appearance of molds on bread and other food, and we use the visual cue to avoid eating contaminated food. Similarly, when we grow bacteria in the lab, we typically look for the formation of colonies. In the field of microbiology, a **colony** is a collection of genetically identical cells that arise from a single cell. This can be any one cell in the environment or in a soil sample. Single cells that give rise to a colony are referred to as **CFUs (colony forming units)**. As a CFU reproduces, the mass of dividing cells eventually becomes large enough that a macroscopic colony becomes visible to the naked eye. Colonies enable us to see into the microscopic world, each colony revealing characteristics that are unique to its species and which, in many cases, can be easily recognized by experienced microbiologists (more on colony morphology in Section 5).

An important step in isolating soil bacteria is obtaining **single colonies**. We know that, by definition, a colony arises from a single CFU. Therefore, we can expect all the cells present in the colony to be of the same genetic composition. By studying bacterial species discretely, we can truly appreciate the diversity of microbes and their unique characteristics.

How do we go from millions of individual cells and species living in soil to single colonies on a solid medium? Even if only 3 in 1,000 bacteria in the soil will grow in the lab, culturing from even a speck of soil will result in tens and hundreds of thousands of colonies growing on the surface of a solid medium all crammed together. The outcome of this will be a nondescript mass of contiguous and overlapping colonies that provides no useful information about the types of microorganisms in the sample. The logical solution would be to spread them out, but how much room do we need to spread out thousands of CFUs so they grow into non-overlapping, distinct colonies?

Microbiologists use dilution methods to reduce the number of cells present in a given sample. We may dilute them in pure water or in salty solutions to keep the cells in an isotonic environment. Once we obtain a desired dilution, which can only be determined on a case-by-case basis, we can spread a small volume of the dilution (in the range of microliters) on a plate. The spreading step separates and disperses cells all over the surface of the plate. Each CFU will be randomly deposited on a spot on the plate where it will proliferate and give rise to a single colony. Colonies will be sparse and separated when the appropriate dilution is achieved (Figure 3-3).

Since colonies represent the unique qualities of the bacteria reproducing in them, we must ensure that they are not contaminated with other species that may obscure our observations (Figure 3-3C). To confirm that a colony contains a single species of bacteria, microbiologists use other spreading methods to further separate cells and observe differences in morphology (more on colony morphology in Section 5). Obtaining **pure cultures**, which each contain a single strain of bacteria, is important for conducting controlled experiments and is the basis of making comparisons, reliable observations, and measures in scientific research. By observing the response of a pure culture to different variables (e.g., type of medium used, temperature, incubation period, technique used, chemicals applied), we can make inferences about the characteristics of the individual cells that constitute it.

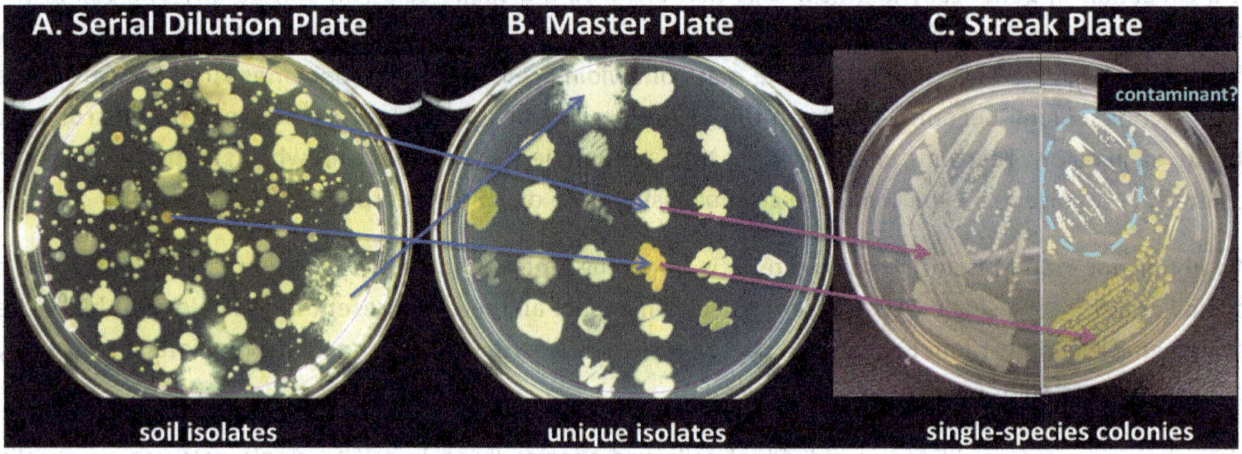

Figure 3-3. (A) A serial dilution plate of a rich soil sample. Bacteria in the soil sample were suspended in sterile water and diluted 10,000-fold (10^{-4}) to obtain a distribution of cells that produces discrete colonies, while representing a wide diversity of bacteria. After incubating the plate 24–72 hours, single colonies can then be picked from the dilution plate and smeared on a new medium plate to create a master plate. (B) The master plate serves as a catalogue of the visibly distinct bacteria that were isolated from a sample. The blue arrows show the transfer of cells from colonies in the dilution plate to a clear spot on the master plate. After 24–72 hours of incubation, the smear grows onto a dense patch of cells that is visible to the naked eye. From this collection of cells, we start new cultures and do further testing. (C) The unique isolates on the master plate are "streaked" onto a new plate to isolate single colonies. A streak of one of the isolates (far right) reveals a second species of bacteria interspersed in the patch. The streaking method allows us to separate them and start pure cultures of each.

Milestones in Microbiology

Selman Waksman coined the term "antibiotic" and pioneered exploration of the prolific antibiotic producers of the soil during the 1940s and 1950s. Along with his student Albert Schatz, he discovered and developed the antibiotic streptomycin, produced by the actinomycete *Streptomyces griseus*. Streptomycin was the cure for tuberculosis. Wakman's discovery, along with the commercialization of penicillin, ushered in the golden age of antibiotics, a time in which the rate of antibiotic discovery was rampant, with dozens of antibiotics being discovered and brought to the market every year. By the beginning of the 1980s, it was believed that antibiotics had eliminated the threat of infectious disease.

Photo source: commons.wikimedia.org

Alma Whiffen, a contemporary of Waksman, was a mycologist who discovered the antifungal agent cycloheximide during the mid-1940s, produced by *S. griseus* (the same bacterium that produced the antibiotic streptomycin). Because of its toxicity to eukaryotic cells (such as human cells), cycloheximide is normally used for research applications, such as inhibiting the growth of fungi in bacterial cultures rather than for treatment of disease in people.

Photo source: commons.wikimedia.org

References

Amann, R. I., Ludwig, W., & Schleifer, K. H. (1995). Phylogenetic identification and in situ detection of individual microbial cells without cultivation. Microbiol Rev, 59:143–169.

Curtis, T. P., Sloan, W. T., & Scannell, J. W. (2002). Estimating prokaryotic diversity and its limits. Proc Natl Acad Sci U S A, 99:10494–10499. doi: 10.1073/pnas.142680199

Schloss, P. D., & Handelsman, J. (2006). Toward a census of bacteria in soil. PLoS Comput Biol, 2:e92. doi: 10.1371/journal.pcbi.0020092

Experiment 3: Find a method to isolate single colonies of bacteria from your soil sample

Biological Questions:

1. What is the significance of single colonies? How can you obtain single colonies on a medium plate from a soil sample?
2. How will you enumerate the culturable microorganisms present in your soil sample?
3. Once you obtain single colonies, can you identify morphologically distinct microorganisms?
4. What would you like to learn about the microorganisms you isolated?

Background:

Objective:

Hypothesis and rationale:

Procedure for obtaining single colonies on plate:

Draw a schematic of the procedure used:

Data and observations:

Interpretations and conclusions:

Notes:

Assignment: Calculating CFU/g soil

Objective:

Sample	Number of colonies per sample				
Media	No dilution	10^{-1}	10^{-2}	10^{-3}	10^{-4}

CFU/g soil calculations:

How does this compare with estimates in the literature?

Notes:

Section 4: Bacteria Are What They Eat, Too

Composition of Living Things

We have considered methods to maximize the number of bacteria in a sample. But did those efforts also capture the number of different species? Will all species present in the soil grow equally well on a single type of media? To answer these questions, we must consider how microorganisms obtain the energy needed to support biological processes and proliferation.

Like all other organisms, microorganisms use energy to produce organic molecules that perform specific biological functions and allow them to grow, proliferate, and engage with their environments. Organic molecules by definition contain carbon (C) atoms, the core elemental building block of life, and are the main constituents of cellular components and energy sources. Some organic molecules can be joined together to form complex macromolecules (see Table 4-1), which perform their own specific functions, or be further modified by reactions within the cell to form secondary metabolites. Organic molecules are largely constructed from only a few elements. Of these elements, carbon (C), hydrogen (H), oxygen (O), nitrogen (N), phosphorus (P), and sulfur (S) are most abundant and required for life. A variety of additional elements are also needed by living cells but in lesser quantities (see Table 4-2).

Organic molecule	**Elemental components**	**Macromolecule**
Amino acids	C, H, O, N, S	Proteins
Nucleic acids	C, H, O, N, P	DNA and RNA
Fatty acids	C, H, O	Lipids
Sugars	C, H, O	Carbohydrates

Table 4-1. Macromolecules and their constituents.

Sodium (Na^+)	Magnesium (Mg^{2+})	Calcium (Ca^{2+})
Iron (Fe^{2+}/Fe^{3+})	Potassium (K^+)	Chloride (Cl^-)

Table 4-2. Trace elements. Table adapted from Slonczewski & Foster (2011)

Classifying Bacteria Based on Nutrition

Bacteria can be classified based on how they obtain carbon and energy from their environments to make biological molecules (Table 4-3). Some organisms produce organic molecules by obtaining or "fixing" carbon from inorganic sources, such as CO_2. They are known as **autotrophs** and are the primary producers in food chains. Photosynthetic organisms, such as plants, algae, and cyanobacteria, are good examples of autotrophs. Conversely, **heterotrophs** obtain carbon from organic molecules formed by other organisms. Animals, protozoa, fungi, and many bacteria and archaea are heterotrophs.

Bacteria can also be classified based on their tolerance for oxygen. Many (but not all) bacteria that live in contact with the atmosphere, which is approximately 21% oxygen gas (O_2) by volume, are **aerobic**, which means they can grow and survive in the presence of O_2. In contrast, O_2 is toxic to some anaerobes, many of which thrive only in strictly O_2-free environments.

Living organisms use oxidizing substances, such as O_2, sulfate (SO_4^{2-}), and pyruvate, as electron acceptors in fueling reactions that oxidize "food" molecules to free usable energy. These processes of respiration (using oxygen as a source of electrons) and fermentation (using organic molecules as a source of electrons) ultimately capture and store released energy in the form of **adenosine triphosphate (ATP)**, the "energy currency" common to all life.

		Energy Source	
		Light (photo-)	**Chemical compounds (chemo-)**
Carbon Source	**Carbon dioxide (auto-)**	**Photoautotrophs** • Plants, algae, cyanobacteria • Green sulfur and purple sulfur bacteria	**Chemoautotrophs** • Hydrogen, sulfur, nitrifying bacteria
	Organic (hetero-)	**Photoheterotrophs** • Green nonsulfur and purple nonsulfur bacteria	**Chemoheterotrophs** • Aerobic respiration: most animals, fungi, and protozoa; many bacteria

Table 4-3: Metabolic types of organisms are categorized on the basis of how they obtain their energy and carbon. Table adapted from Bauman, Microbiology 1st ed., Figure 6.1

Media and Culture Conditions

In the lab, microbes are grown on many different types of media that contain mixes of essential nutrients suspended in water (e.g., broths and infusions). Some media are solidified with agar, gelatin, or another jelly-like substance. Media ingredients range from dehydrated extracts of plant, animal, or fungal material to synthetic compounds made in the lab. Media can be enriched with a variety of organic molecules, macromolecules, and vitamins to support the growth of **fastidious** bacteria, which have complex nutritional requirements. They can also be minimal, composed mainly of simple sugars, salts, and water. Over the years, microbiologists have developed standard media formulations that can be purchased from commercial vendors and used for many different applications.

Other **physical parameters**, such as light exposure, temperature, and tonicity, can drastically affect the assortment of microbes that grow in culture. The composition and concentration of ions and molecules in a bacterium's environment not only influence its nutrition, but also the overall stability of its cells. A medium with inadequate **electrolytes** may represent a hypotonic environment that will cause cells to swell up and burst. Conversely, a hypertonic medium with an excessively high concentration of solutes may cause cells to lose water, shrivel, and die. Similarly, many bacteria cannot survive outside a relatively narrow pH range.

Growth on a plate indicates that the bacteria are living within an acceptable range of conditions that allows them to survive and proliferate. It does not mean that all of their needs are met. Like all living things, bacteria are hard-wired to respond to stimuli in their environments. Cellular processes, such as gene expression, are mediated by factors and complex pathways that scientists do not fully understand and cannot account for in the laboratory. For example, many scientists are trying to stimulate the expression of **silent genes** encoding production of potentially novel secondary metabolites in various species of the genus *Streptomyces*.

Finding the right combination of conditions and stimuli that promote the growth of certain bacteria or induce the expression of certain genes is perhaps the biggest obstacle to unraveling the biodiversity and biosynthetic potential of microbes. Increasingly, research shows that many bacteria are dependent on microbial communities where they exchange signals, genetic material, nutrients, and other stimuli that trigger specific responses in cells. There are more variables than any experiment can test, but the more ways we isolate and grow bacteria, the more we can learn about the organisms in a sample. Use your imagination.

References

Bauman, R. (2003) *Microbiology* (1st ed.) San Francisco: Benjamin Cummings.

Mukhopadhyaya, P. N., Deb, C., Lahiri, C., & Roy, P. (2000). A *soxA* gene, encoding a diheme cytochrome C, and a *sox* locus, essential for sulfur oxidation in a new sulfur lithotrophic bacterium. J Bacteriol, 182:4278-4287.

Slonczewski, J., & Foster, J. W. (2011). *Microbiology: An Evolving Science* (2nd ed.). New York: W.W. Norton & Co.

Living on a Bacterial Planet

The long chains (filaments) of the sulfur-oxidizing bacterium *Beggiatoa*, seen under the microscope under high magnification (notice the scale in the picture). This microorganism is an example of a lithotroph, which derives its nutrition from inorganic molecules, such as reduced inorganic compounds and minerals typically found in rocks, hence the name, "rock eaters." Lithotrophs, composed of both bacteria and archaea, live in extreme environments ranging from bedrock to deep hydrothermal vents. The sulfur-oxidizing bacterium *Beggiatoa* was the first example of a lithotrophic bacterium discovered in microbial mats near hydrothermal vents and sulfur-rich environments in the ocean bottom. *Beggiatoa* oxidizes sulfur compounds such as sulfide (S^{2-}) as their energy source and carbon dioxide (CO_2) as a carbon source (Mukhopadhyaya, Deb, Lahiri, & Roy, 2000)

Photo source: https://alchetron.com/Acidophile

Experiment 4: Choose your own media and culture conditions

Biological Questions:

1. How can we optimize media and culture conditions to favor the growth of diverse microorganisms in your soil sample?
2. What medium and culture conditions do you choose and why?
3. How do your conditions of choice compare with your isolates' natural habitat?

Background:

Objective:

Medium of choice and rationale:

Culture condition choices:

Temperature:

Oxygen level:

Other:

Hypothesis and rationale:

Tiny Earth

studentsourcing antibiotic discovery

Experimental design and techniques used:

Control and treatments:

Data and observations:

Interpretations and conclusions:

Notes:

Section 5: Solid Versus Liquid Cultures

Colony Morphology

Since bacteria were first cultured on solid media, microbiologists have noticed that colonies differ in color, texture, shape, and margin or edges. Works like *Bergey's Manual of Systematic Bacteriology* contain exhaustive descriptions of the physical and biochemical attributes of bacteria and have helped microbiologists identify and classify bacteria for nearly a century. **Colony morphology** is one of the first characteristics used to distinguish a particular organism from other species of bacteria. So it is important that we describe the physical characteristics of our microbes when isolated from the environment (Figure 5-1).

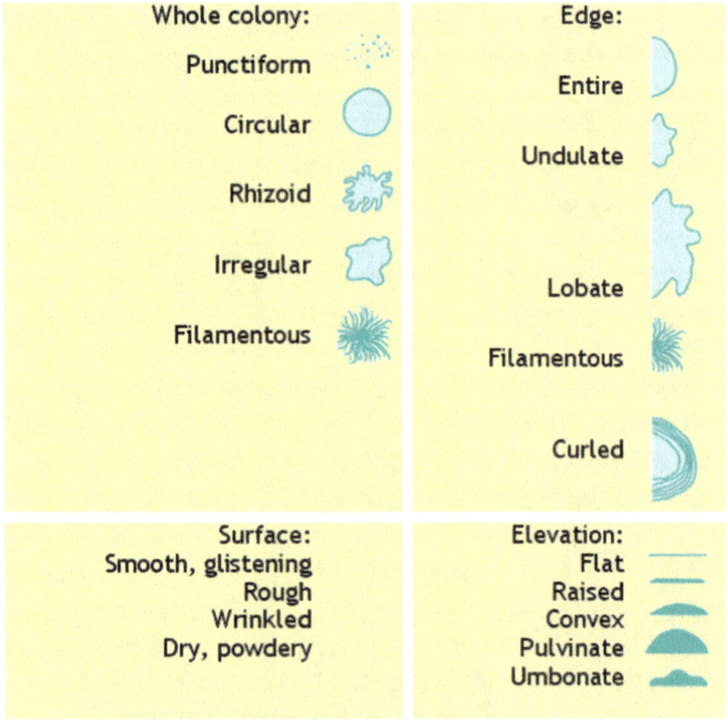

Figure 5-1. The shapes, edges, and elevations of bacterial colonies. This is a useful guide for describing our soil microbes and distinguishing them from others. Photo source: commons.wikimedia.org

Bacteria express characteristic colony morphologies. For example, *Lysobacter* and *Pseudomonas* colonies are typically slimy and irregular in shape and may fluoresce under UV light (Figure 5-2A). *Streptomyces* colonies have bold colors and develop a white fuzzy surface, which contains spores (Figure 5-2B). *Bacillus* colonies may sprawl thinly over the agar and typically show relatively dull, bland colors.

Unfortunately, in the prokaryotic world, morphology does not reliably indicate relatedness and is therefore inadequate to identify an unknown organism. Ultimately, morphology only provides us with a hint, and morphological observation is only the first step of many when attempting to identify a bacterial isolate. To assign a family, genus, or species, we need much more information acquired through further testing. In future sections, we will explore more reliable and modern approaches to bacterial identification.

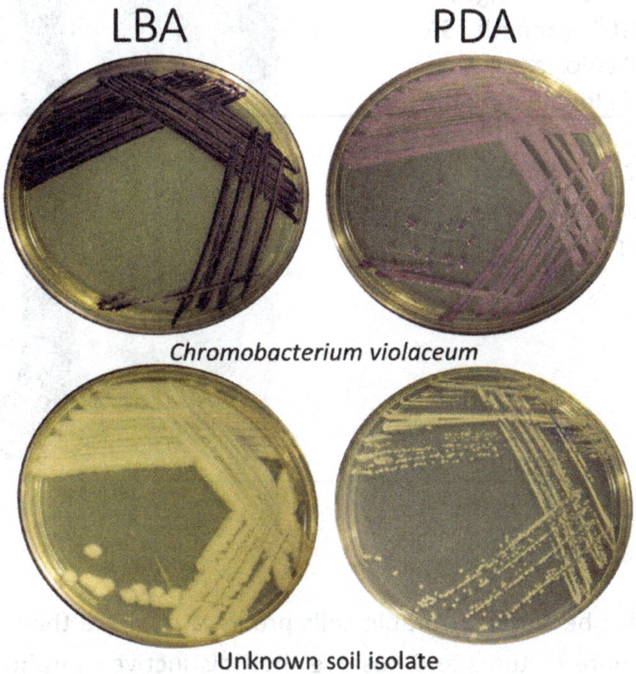

Figure 5-2 Bacteria display a variety of colony colors and morphologies on different media. Top: *Chromobacterium violaceum* produces smooth colonies with a dark violet metallic sheen (due to violacein production) on Luria-Bertani agar (LBA) and pale violet colonies on potato dextrose agar (PDA). Bottom: Unknown soil isolate producing non-pigmented irregular colonies on LBA and colonies with smooth margins on PDA. Photo source: Manuel Fernando Garavito Diago

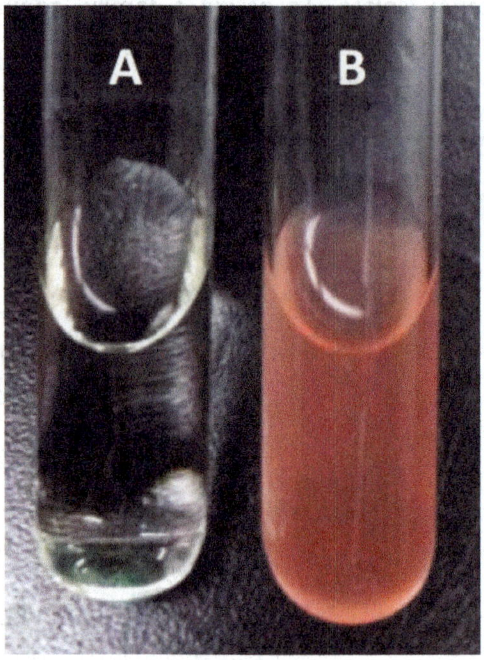

Figure 5-3 [Right]. Two test tubes each containing the bacterial growth medium 1/10th-strength Tryptic Soy Broth (TSA). (A) 1/10th-strength TSA typically looks transparent and colorless in the absence of bacterial growth. (B) Inoculating the medium with a colony of *Lysobacter antibioticus* produces a pink culture after 24–36 hours of growth at 28°C. This microorganism retains its distinctive characteristic—the pigment—in both liquid and solid culture.

Growth Phases

Solid cultures are advantageous because culturable cells proliferate where they were deposited, giving us the opportunity to isolate pure cultures and observe their distinctive morphologies. The opposite is true for liquid media. The fluid nature of liquid cultures allows cells to move randomly throughout the medium. Therefore, to effectively study bacteria in liquid culture, we typically inoculate the medium from a known pure culture, such as a single colony on a solid medium. By doing so, we are assured that the liquid culture remains a pure culture as long as appropriate aseptic technique has been practiced.

Liquid cultures are particularly suitable for studying bacterial growth, or **growth kinetics**, and biochemistry at different growth phases (Slonczewski & Foster, 2011, p. 126). Unlike a solid culture in which cells are normally in direct contact with one another and/or their substrate, cells in liquid culture are in a **planktonic,** or free-floating, state. This allows them to move throughout the medium and, if properly shaken during growth, form uniformly mixed cultures. Many bacteria require their liquid environment to be continually aerated and mixed. Therefore, liquid cultures are often grown with the assistance of shaker platforms. Caution must be exercised, however— although good for bacteria that can thrive in a planktonic state, shaking is not advisable for organisms that prefer a solid substrate. Many soil bacteria are not suited for planktonic growth and therefore when transferred from solid to liquid culture, there is a decline in the number of viable cells.

Bacterial growth can be approximated by measuring how much light of a particular wavelength (600 nm) is absorbed by a liquid culture over time. This is done with an instrument called a **spectrophotometer**, which measures absorption as **optical density units (ODs)**. As cell concentration increases, the liquid culture becomes more cloudy or turbid. As the **turbidity** of the culture increases, the OD value reported by the spectrophotometer will also increase. The mathematical relationship between OD and cell concentration can be used to determine rates of bacterial growth over time. When OD is plotted versus incubation time, we obtain a **standard growth curve** (Figure 5-4), which is a useful model for keeping track of how a population of cells in liquid culture changes over time or predicting future behavior of the same isolate. The model shows that bacteria undergo various phases of growth in liquid culture. Growth phases are indicative of what their cells are doing physiologically and metabolically.

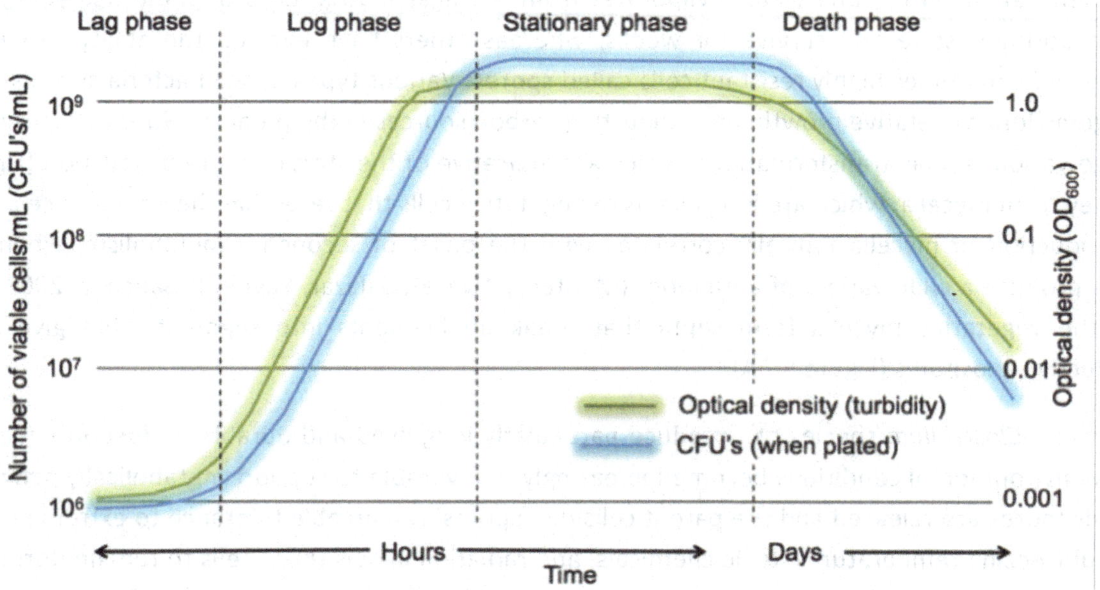

Figure 5-4. Standard bacterial growth curve and growth phases of *E. coli* in liquid culture. Graph adapted from Wessner, Dupont, & Charles (2013)

When cells are first transferred into a new medium, they will enter a period known as the **lag phase** during which very little growth occurs. During this period, the cells taking cues from the environment and making metabolic adjustments, altering patterns of gene expression to accommodate the new conditions. Once this process has completed, the cells will begin to divide rapidly. Microbiologists typically work with bacteria in the **logarithmic ("log") phase**, where cells are found reproducing exponentially. Bacteria in this phase are highly metabolically active, consuming nutrients in the medium and synthesizing macromolecules to make new cells.

Secondary metabolites, such as pigments and antibiotics, are normally produced toward the end of the log phase as cell concentration peaks and nutrients become depleted. Bacteria in the genus *Streptomyces* have been found to enter into secondary metabolism after they cease to produce essential cellular components, such as DNA (Bibb, 1996). Therefore, for this microorganism, antibiotic production occurs in a growth-phase-dependent manner, which is in turn influenced by nutrient availability in the medium. As nutrients are depleted and waste products accumulate, growth slows such that the number of cells dividing is approximately equal to those dying. This is called **stationary phase**, which can last hours to days depending on the cells' response to a lack of nutrients and an increasingly toxic environment. Eventually, the number of dying cells surpasses the number of living and reproducing cells, and the culture enters the **death phase.**

Spores

Many markers of bacterial growth phases are also visible in solid culture. Over time, colonies stop spreading on the agar and may even begin to shrivel. This occurs as nutrients are depleted, waste and toxic chemicals build up, and water evaporates from the agar matrix. Cells of some bacterial species enter a dormant state and survive for weeks, whereas others have evolved the ability to generate inactive and sometimes highly resistant cells called **spores**. Various types of soil bacteria sporulate after undergoing long vegetative growth and exhausting carbon sources in the medium. *Streptomyces* species undergo various visible transformations, which are indicative of the start of spore production (Figure 5-2B). They form mycelia, which are long chains of vegetative cells that resemble the appearance of fungi. The production of mycelia typically correlates with the onset of secondary metabolism, when these species produce a wide variety of antibiotics (Manteca, Alvarez, Salazar, Yague, & Sanchez, 2008). Soon after, the vegetative mycelia form septa that break up filaments into segments that give rise to reproductive exospores (Figure 5-5A).

Bacillus and *Clostridium* species can produce particularly long-lived and durable **endospores** (Figure 5-5B). As environmental conditions become increasingly unfavorable to support metabolically active cells, the endospores are released and the parent cells die. Spores' remarkable tolerance to extreme dryness, heat, subfreezing temperatures, toxic chemicals, and radiation allows these cells to remain dormant for long periods of time. There have been reliable reports of the successful germination of 100,000-year-old endospores, and even (somewhat controversial) reports of 25-40 million-year-old spore germination (Nicholson et al., 2000). Avoiding contamination with long-lived endospores is the primary reason that lab supplies must be autoclaved in order to achieve sterilization. Autoclaving involves exposure to pressurized steam at 15 PSI and 121°C for a suitable period of time. (15 minutes is appropriate for small volumes.)

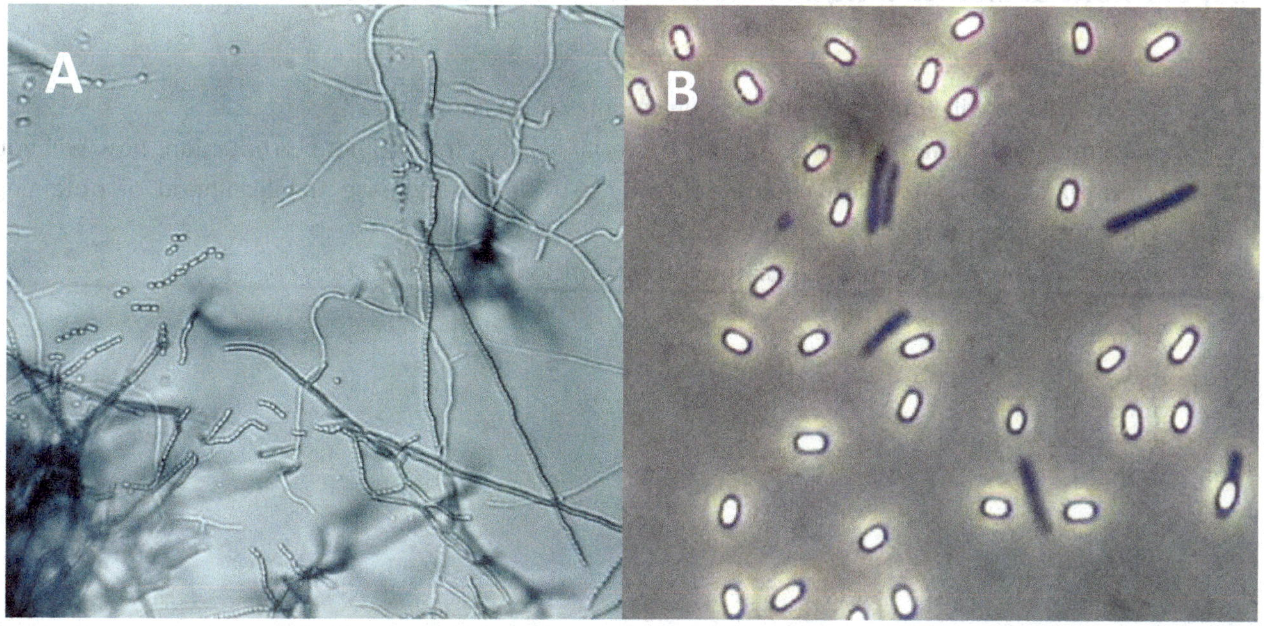

Figure 5-5. Two types of bacterial spores observed under phase contrast at 400X [left] and 1000X [right] magnification. Left: *Streptomyces* sp. mycelial filaments and long chains of exospores. Right: *Bacillus* sp. endospores appear as bright, dense structures under the microscope. Photo sources: [Left] commons.wikimedia.org (credit to CDC Public Health Image Library); [Right] commons.wikimedia.org

References

Bergey's Manual of Systematic Bacteriology (2nd ed.) 2005. New York: Springer.

Bibb, M. (1996). The regulation of antibiotic production in *Streptomyces coelicolor* A3(2). Microbiology 142:1335–1344.

Manteca, A., Alvarez, R., Salazar, N., Yague, P., & Sanchez, J. (2008). Mycelium differentiation and antibiotic production in submerged cultures of *Streptomyces coelicolor*. Appl Environ Microbiol 74:3877–3886. doi: 10.1128/AEM.02715-07.

Nicholson, W. L., Munakata, N., Horneck, G., Melosh, H. J., & Setlow, P. (2000). Resistance of Bacillus endospores to extreme terrestrial and extraterrestrial environments. Microbiol Mol Biol Rev 64:548–572.

Slonczewski, J., & Foster, J. W. (2011). *Microbiology: An Evolving Science* (2nd ed.). New York: W.W. Norton.

Wessner, D. R., Dupont, C., & Charles, T. (2013). *Microbiology* New York: Wiley Pub.

Experiment 5: Isolate unique colonies to test for antibiotic production

Biological Questions:

1. What criteria will you use to pick unique colonies from your plates?
2. Assuming you cannot test ALL of your bacterial isolates for antibiotic production, how will you prioritize the isolates you choose to study in order to increase the likelihood of obtaining antibiotic producers?
3. What are the qualities of the isolates you have picked for further study?

Background:

Objective:

Hypothesis and rationale:

Protocol:

Results:

Number of isolates picked:

Characteristics of isolates:

Interpretations and conclusions:

Notes:

Section 6: Meet the ESKAPE Pathogens

Understanding Infectious Disease

Since ancient times, many diseases, including cholera and the bubonic plague, were attributed to miasmas, or "bad air." People believed that pests and parasites associated with plagues, decay, and spoilage could arise from inanimate, or nonliving, matter. This idea – known as **spontaneous generation** – along with the miasma theory greatly hindered preindustrial society from making the connection between microbes and disease and prevented development of effective strategies for prevention and control of disease outbreaks.

In the mid-1800s, **John Snow** helped dispel some of these misconceptions by describing the mode of transmission of cholera. He discovered the source of a particular cholera outbreak in London – a communal water pump in a dirty, densely populated neighborhood. The water source had been contaminated by sewage runoff from a house inhabited by cholera victims. The outbreak killed hundreds of people who drank water from the pump. Snow's detailed epidemiological data led to the disabling of the pump and increased awareness of water quality, public hygiene, and sanitation.

During the same time period, **Louis Pasteur**, a French chemist and microbiologist, helped disprove the theory of spontaneous generation. In an elegant experiment, he showed that sterilized broth, similar to the growth medium we use in the lab, does not spontaneously generate microorganisms from nonliving components. Instead, microbial growth originates as contamination introduced from the surrounding air. This finding was particularly important in an era when food spoilage was common, leading to great losses in disease-ridden industrial cities where food was already scarce. Pasteur realized that the key to preservation was to reduce the number of bacteria present in foods and beverages by applying heat in a process that came to be known as **pasteurization**.

Pasteur's revolutionary work supported the **germ theory of disease**, which states that microorganisms like bacteria are the cause of infectious disease. At the same time, scientists were also becoming aware that microorganisms were virtually everywhere and had useful applications as well. Yeast were shown to be responsible for the fermentation of grape juice to produce wine, and bacterial fermentation was found to convert milk to yogurt.

Work by the German physician **Robert Koch** provided the means to identify **pathogens**, or disease-causing microorganisms. While performing experiments to identify the agents responsible for tuberculosis and anthrax, Koch established a series of criteria, or "postulates," that could be applied to identify pathogens responsible for other infectious diseases as well:

1. *The microbe is always present in diseased individuals but not healthy individuals.*
2. *The microbe is isolated from the infected individual and grown in a pure culture.*
3. *When the microbe in pure culture is introduced into a healthy individual, the same disease symptoms occur.*
4. *The microbe reisolated in pure culture from the diseased individual has the same characteristics as the original.*

This process might seem second nature to us today, but in the late 1800s, it was revolutionary. Koch successfully identified *Bacillus anthracis* and *Mycobacterium tuberculosis* as the causative agents of anthrax and tuberculosis, respectively, and was awarded the Nobel Prize in Physiology or Medicine in 1905. Koch's simple and logical criteria continue to help modern clinical microbiologists identify pathogens. Today, the link between microbes and disease is well understood; this knowledge has enabled development of effective measures to prevent and treat infectious disease around the world. Nevertheless, humanity remains at risk from diseases caused by microorganisms.

Pathogens that were once believed to have been conquered during the golden age of antibiotics (see Section 7) are now reemerging in drug-resistant forms, and they are doing so at an accelerating rate. Consequently, we are effectively running out of treatment options. Hospitals have become havens for multi-drug resistant "superbugs," and the widespread misuse of antibiotics in households and in agricultural practices around the world means that drug-resistant pathogens are emerging as a threat in our homes and environment. The process by which pathogens acquire resistance is only partially understood and is an important area of ongoing research (see Section 10), but the trends we see today point to a grim future if we do not find effective solutions based on drug deployment, hygiene, or other practices.

Choosing Test Organisms

- *Since antibiotics do not affect all bacteria equally, how will you decide which organism to test your soil microbe against?*

As we isolate individual microbes, we hope that some will produce antibiotics at detectable levels. Initially, for practical reasons, researchers assay these microbes for the ability to inhibit the growth of only a single or small number of test organisms. Microbes that show activity may later be screened against a broader panel of test organisms.

- *Does the ability to inhibit the growth of one test organism indicate that the antibiotic will inhibit all bacteria?*

Because prokaryotes differ in their anatomical and physiological traits, different organisms may be more or less susceptible to the effects of any particular antibiotic. Some antibiotics are **broad spectrum**, meaning that they affect a wide range of bacteria. Other antibiotics have a **narrow spectrum** of activity. One anatomical feature that plays a significant role in the susceptibility of a microbe to a particular antibiotic is its cell wall composition (discussed in Section 8).

- *How can we detect production of antibiotics? How do we distinguish the producers from the nonproducers?*

In Section 7, we will conduct experiments to screen our isolates for their ability to produce antibiotics. But first, we must decide which microbe to test our isolates against.

The ESKAPE Pathogens

When choosing a test organism, a logical choice is an organism that is clinically relevant – i.e., a human pathogen. There are six organisms that are today considered to be major threats, not because they cause the most devastating illnesses but because they comprise the majority of antibiotic-resistant infections seen in health care settings. They are ***E**nterococcus faecium*, ***S**taphylococcus aureus*, ***K**lebsiella pneumoniae*, ***A**cinetobacter baumannii*, ***P**seudomonas aeruginosa*, and several species of ***E**nterobacter* (**ESKAPE**).

The leading author of a 2009 report on Clinical Infectious Diseases, Helen Boucher, MD, stated:

> *These [ESKAPE pathogens] are among the biggest threats infectious disease physicians face today... We desperately need new drugs to fight them. But we also need cooperation among industry, academia, and government to create a sustainable R&D [research and development] infrastructure that will fill the pipeline to meet today's needs and keep it filled with drugs that tackle tomorrow's infectious diseases threats.*

ESKAPE Pathogen	Safe Relative
Enterococcus faecium	*Enterococcus raffinosus*
Staphylococcus aureus	*Staphylococcus epidermidis*
Klebsiella pneumoniae	*Escherichia coli*
Acinetobacter baumannii	*Acinetobacter baylyi*
Pseudomonas aeruginosa	*Pseudomonas putida*
Enterobacter species	*Enterobacter aerogenes*

Table 6-1. The ESKAPE pathogens and their safe relatives, used as tester organisms by **Tiny Earth**

Although it would be ideal to test the ability of our isolates to inhibit the growth of all or at least a subset of these organisms, it is not possible to do so in a teaching laboratory due to their potential to cause human disease. However, we can take advantage of evolution and test our isolates against closely related organisms that do not pose a health risk. These "safe relatives" have many of the same anatomical and physiological features as their respective ESKAPE counterparts, yet they are safe enough to be studied in teaching laboratories as long as basic biosafety protocols are followed. In addition to the six safe relatives listed above, students often test their soil isolates against *Mycobacterium smegmatis,* which is a safe relative of the causal agent of tuberculosis, *Mycobacterium tuberculosis*.

Not All Bacteria Are Harmful

Although pathogenic bacteria get most of the attention (for good reason), it's important to recognize that humans are constantly associated with bacteria in our bodies, in the foods we eat, and in the environment. If you were to examine every cell in and on your body, you would find that the majority of them are not actually human! The members of the **human microbiome**, the microbes that reside in and on our bodies, outnumber our own cells 2:1 (Sender et al., 2016). Research has shown that our cells are "constantly bathed" in bacterial metabolites (Slonczewski & Foster, 2011). Therefore, there can be no doubt that microbes intimately and substantially influence our lives and our health.

The human gut, which houses 100 billion bacteria per cm^3, teems with bacterial activity that is responsible for 15–20 percent of our caloric intake (Slonczewski & Foster, 2011, p. 868). This **consortium** of microorganisms is composed of organisms such as *Bacteroides, Clostridium,* and *Escherichia* species, among many others. Disruption of the gut microbiota can cause many different metabolic and physiological disorders.

Various Gram-positive bacteria, such as *Staphylococcus* and *Bacillus*, colonize our skin and are also part of the normal human microbiome. They are not simply passive bystanders, but are an important part of our first-defenses against infection and are active inducers of proper immune function. In an experiment by Belkaid et al., mice with *Staphylococcus epidermidis* on their skin were found to have a better immune response than germ-free mice (Naik et al., 2012).

Probiotics, microbes that are consumed to improve health, are found in many fermented products, including yogurt. Pharmacologist John Cryan carried out experiments to test the neurological effects of administering doses of probiotics to mice. He found that certain probiotics had impressive effects on mouse stress and anxiety levels. Mice treated with probiotics were more willing to walk out into the open and take action to protect themselves from a perceived threat of drowning, behaviors suggestive of "happy" mice with lower anxiety levels (Bravo et al., 2011). Other research has suggested that residents of the mammalian gastrointestinal tract influence asthma, autism, colitis, cancer, obesity, and diabetes. These findings are revolutionizing the way we think about microbes and could have exciting applications in humans.

Biotechnology is the application of organisms and their products to address human needs. We have used yeast and bacteria to ferment foods and beverages and improve crop health for millennia. Bacterial products are now found in household items and cosmetics and used in medical procedures. We use bacterial lipases and proteases as detergent additives and toxins like the botulinum toxin (known commercially as Botox) to reduce wrinkles and treat certain diseases. Furthermore, we use microbes as basic tools in bioengineering and chemical manufacturing. Human genes cloned into bacteria and yeast serve as pharmaceutical factories that produce valuable proteins, such as insulin (Slonczewski & Foster, 2011, p. 35). Bacteria can be used to break down pollutants and detoxify contaminated environments, a process known as bioremediation. In 2010, in the Gulf of Mexico, hydrocarbon-consuming microbes helped to degrade much of the Deepwater Horizon oil spill (Biello, 2010). Microbes can even help recycle elements trapped in landfill garbage and return them to the biosphere.

While a subset of microbes are pathogenic to humans, animals, or plants, the benefit we realize from the vast majority of microbes seems almost endless. In the following sections, we will learn about microbial compounds with pharmaceutical activity and the roles they may play in Nature.

References

Biello, D. (2010) Meet the microbes eating the gulf oil spill. Scientific American August 18, 2010. <http://www.scientificamerican.com/article.cfm?id=gulf-oil-eating-microbes-slide-show>

Boucher, H. W., Talbot, G. H., Bradley, J. S., Edwards, J. E., Gilbert, D., Rice, L. B., Scheld, M., Spellberg, B., & Bartlett, J. (2009) Bad bugs, no drugs: no ESKAPE! An update from the Infectious Diseases Society of America. Clin Infect Dis 48:1–12.

Bravo, J. A., Forsythe, P., Chew, M. V., Escaravage, E., Savignac, H. M., Dinan, T. G., Bienenstock, J., & Cryan, J. F. (2011) Ingestion of *Lactobacillus* strain regulates emotional behavior and central GABA receptor expression in a mouse via the vagus nerve. Proc Natl Acad Sci U S A 108:16050–16055. doi: 10.1073/pnas.1102999108.

Naik, S., Bouladoux, N., Wilhelm, C., Molloy, M. J., Salcedo, R., Kastenmuller, W., Deming, C., Quinones, M., Koo, L., Conlan, S., Spencer, S., Hall, J. A., Dzutsey, A., Kong, H., Campbell, D. J., Trinchieri, G., & Segre, J. A., Belkaid, Y. (2012) Compartmentalized control of skin immunity by resident commensals. Science 337:1115–1119. doi: 10.1126/science.1225152

Sender, R., Fuchs, S., Milo, R. (2016) Revised estimates for the number of human and bacteria cells in the body. PLoS Biol 14(8): e1002533. https://doi.org/10.1371/journal.pbio.1002533

Slonczewski, J., & Foster, J. W. (2011) *Microbiology: An Evolving Science* (2nd ed.). New York: W.W. Norton.

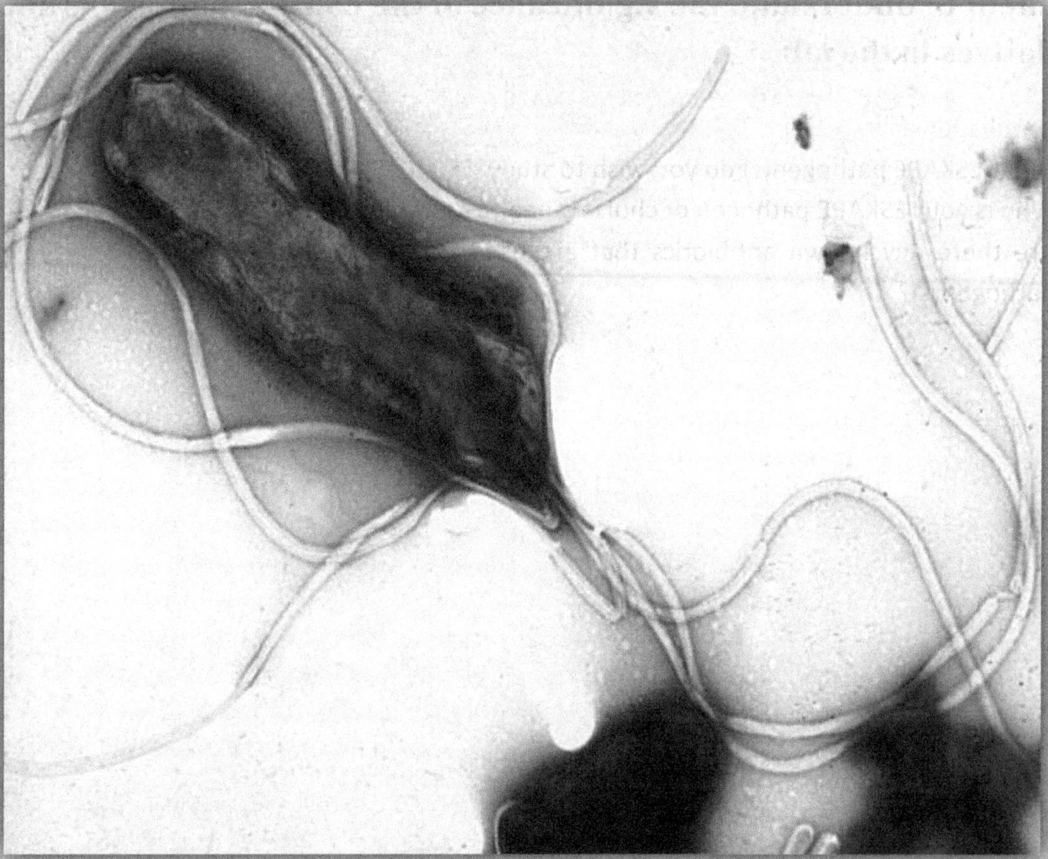

Figure 6-1. This electron micrograph shows *Helicobacter pylori*, a resident of the stomach, with multiple flagella that the bacterium uses to move in the mucous layer of the stomach lining. Photo source: commons.wikimedia.org

Milestones in Microbiology

The Australian scientist Barry Marshall discovered that peptic ulcers and gastritis are caused by bacteria living in the stomach, which was generally believed to be too acidic for bacteria to survive. His inability to isolate and culture the bacterium, *Helicobacter pylori*, damaged his credibility in the scientific community and sustained commonly held myths about the stomach and ulcers. In a courageous effort to validate his findings, Marshall drank a culture of the infectious bacterium and developed symptoms of gastritis after a few days. An endoscopy and biopsy demonstrated that the bacterium was in fact present on the stomach lining, which was inflamed. After antibiotic treatment, the symptoms subsided, providing compelling evidence that the *H. pylori* was, in fact, the cause of ulcers. Marshall conducted the experiment in the 1980s and received the 2005 Nobel Prize for his work, which revolutionized views about the extreme conditions in which bacteria can grow, the assumption that all bacteria can be cultured, and the standard treatment for ulcers.

Experiment 6: Understand the significance of the ESKAPE pathogens and using safe relatives in the lab

Biological Questions:

1. What ESKAPE pathogen(s) do you wish to study?
2. Why is your ESKAPE pathogen of choice a health concern?
3. Are there any known antibiotics that are used to treat an infection caused by your ESKAPE pathogen(s)?

Background:

Objective:

ESKAPE Pathogen(s) of Choice:

Reasons for Choosing ESKAPE Pathogen:

General Cellular and Morphological Characteristics of the Organism:

Clinical Importance and Prevalence:

Infection (How does the infection occur and where is it localized?):

Pathology (What disease is caused? What are the symptoms?):

Ineffective Antibiotics (Antibiotics to which the organism has acquired resistance):

Effective Antibiotics (Antibiotics known to inhibit the organism):

Corresponding Safe Relative:

Notes:

Section 7: Antibiotic Discovery, Structure, & Targets

The Father of Antibiotics

Alexander Fleming is famous as the "accidental" discoverer of **penicillin** – the first antibiotic, which revolutionized the world of medicine. But how "accidental" was his discovery really? Was Fleming really the first to observe this type of amazing microbial activity? In reality, this bacteriologist made his discovery because he already knew what he was looking for. Years of practice in his field had provided him with keen skills of observation and inquiry, enabling him to turn a small observation into a find with monumental implications.

Throughout the 1800s, highly corrosive **antiseptics** were being used to treat severe infections, especially those suffered by soldiers wounded in wars. These antiseptics were damaging to human tissue as well as to the human immune system. Most casualties of war actually resulted from the infections and treatments given after the battles. This situation was referred to by scientists and medical professionals as the **antiseptic dilemma**.

During the early 1900s, Fleming began his quest for an alternative to antiseptics – a chemical that could selectively kill bacteria without harming host cells. This idea resonated with Paul Ehrlich's **magic bullet** hypothesis, which stated that if certain chemicals, such as dyes, could selectively bind to specific bacteria, then there should be compounds that selectively inhibit growth as well. Ehrlich went on to develop the first human-made antimicrobial, salvarsan, in 1909. Salvarsan was effective in syphilis treatment but was later found to be too toxic for human use (Amyes, 2001), making it a great example of the antiseptic dilemma.

It was not until 1928 that Fleming came across his rather famous messy stack of plates awaiting disposal. As Fleming examined the plates, he noticed that one was contaminated. Moldy plates were not uncommon in the lab; however, there was something unique about this plate: *Staphylococcus* cells were not growing along the perimeter of the mold, forming a **zone of inhibition** (Fleming, 2001)! Fleming was intrigued by this observation and eventually showed that the zone of inhibition was caused by a compound that he named penicillin after the fungus that produces it, *Penicillium notatum*.

Characteristics of antibiotics:

- Therapeutic agents
- Selectively toxic toward microbes
- Generally antibacterial
- Small molecules (secondary metabolites)
- Originally referred to compounds produced by microorganisms, but today the term is used also for synthetic antimicrobials

To test the effectiveness of penicillin, Fleming performed various activity assays against common pathogens. Penicillin inhibited the growth of Gram-positive bacteria associated with scarlet fever, pneumonia, gonorrhea, meningitis, and strep throat, while leaving eukaryotic cells unharmed (Brown, 2005). It exhibited the qualities of the miracle drug Fleming had been pursuing for years. However, he was unable to purify the compound in an active and stable form, making it impossible to test *in vivo*. He could not test penicillin on humans at concentrations that would reach and have impact at the site of infection without exposing test subjects to extraneous compounds also present in the mold culture.

Work on penicillin lay dormant for ten years until pharmacologist Howard Florey and biochemist Ernst Chain came across Fleming's paper. Their combined knowledge of chemistry allowed them to purify the active compound and test it on infectious bacteria *in vivo*, proving its therapeutic ability in humans (Amyes, 2001). Their work, coupled with the extraordinarily high-producing strain discovered by Mary Hunt and Kenneth Raper, enabled penicillin to be mass produced and administered to patients orally. Penicillin was commercialized in time to reach wounded soldiers on the battlefields of World War II and significantly contributed to the Allied victory and making World War II the first war in America's history in which more soldiers died of the direct effects of bullets and bombs than of infections. The collaborative development of the first natural product antibiotic won Fleming, Florey, and Chain the Nobel Prize in Physiology or Medicine in 1945.

Milestones in Microbiology

The development of penicillin led to a worldwide search for a strain of the *Penicillium* mold that would produce copious amounts of the antibiotic and grow in liquid culture. Mary Hunt (popularly known as Moldy Mary), a technician at the USDA lab in Peoria, Illinois run by Dr. Kenneth Raper, scoured produce in local markets in search of the strain and came across a cantaloupe carrying a mold that produced the highest concentrations of penicillin ever observed in a strain of *Penicillium.* The strain was improved and grown in deep-vat, submerged conditions, enabling the antibiotic to be mass-produced and sent to the battlefront of World War II.

Photo source: http://www.peoriahistoricalsociety.org/!/Exhibits-PenicillinMoldyMary

Milestones in Microbiology

It is worth noting that Fleming was not the first one to make these observations. Texts from Ancient Egypt describe the use of moldy bread to treat pustular eruptions on the scalp (Böttcher, 1959), while the Bible mentions the curative value of hyssop. It was on this plant (photo above) that the Swedish naturalist Westling found *Penicillium notatum* thousands of years later (Brown, 2005). Although Fleming himself coined the term "penicillin," an inscription on a therapeutic ointment with the word "PENICILLE," found in an Ancient Roman site, perplexed archeologists many years later after this drug became a common household name (Böttcher). Yet, through all these years, people had exploited these molds for their curative value without understanding the chemistry and biology underpinning the effects. Many scientists attempted to uncover its inhibitory properties in the 1800s after Louis Pasteur discovered the role of bacteria in infection. Joseph Lister noticed that the *Penicillium* mold inhibited the reproduction of bacteria ten years before Fleming was born and, ten years after, another scientist attributed this inhibition to "the secretion of a specific substance," with no further clarification as to what was secreting it or what the substance was. French scientist Ernest Duchesne even injected the culture broth into mice infected with typhoid, and they were cured soon after. Nevertheless, none of these scientists identified the chemical or pursued the research to a solid conclusion.

This story is quite revealing about the importance of record keeping, serendipity, and persistence in discovery!

Photo source: commons.wikimedia.org

Antibiotic Structure

Penicillin was special due to its ability to kill bacteria without the toxic side effects of antiseptics, which lacked selectivity and harmed human and bacterial cells alike. In 1942, Selman Waksman coined the term "antibiotic" to describe any small molecule produced by a microbe that kills or inhibits the growth of other microbes (Clardy, Fischbach, & Currie, 2009). "Small molecules" refers to the fact that antibiotics are smaller than the macromolecules that constitute a cell. Antibiotics vary in their structures (Figure 7-1), falling into several chemical groups, or **classes**, of antibiotics. Entirely new classes of antibiotics are even more challenging to discover than new antibiotics themselves. A common class of antibiotic is the β-lactams, which share a four-member ring structure, which is the core of β-lactams like penicillin, that enables them to inhibit bacterial cell wall synthesis (more on targets below) (Figure 7-1). One of the most recent classes of antibiotics was discovered in 2004, nearly 300 meters below the surface in the Sea of Japan. The antibiotic class was aptly named "abyssomicins" after the abyss of the deep ocean. These are polycyclic antibiotics produced by marine bacteria in the genus *Verrucosispora*. Abyssomicin C inhibits the growth of vancomycin- and methicillin-resistant strains of *Staphylococcus aureus* by blocking their ability to produce folic acid (Nicolaou, Harrison, & Chen, 2009).

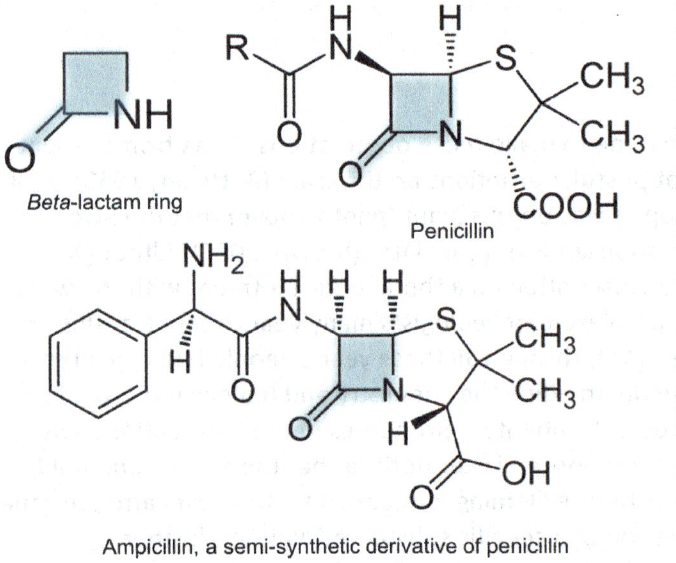

Figure 7-1. Members of the β-lactam class of antibiotics, such as penicillin and ampicillin, all share a common core structure: a four-member β-lactam ring. This structure makes β-lactam antibiotics able to inhibit bacterial cell wall synthesis.

Strictly speaking, true antibiotics are antimicrobial compounds produced by microorganisms. This category of antimicrobials includes common drugs, such as penicillin, streptomycin, and chloramphenicol, which were originally isolated from living microorganisms. Microbial products make up 60-80% of all antimicrobials; the rest are generally synthetically derived from known chemical scaffolds and developed by chemists. Today, "antibiotic" is used to refer to compounds produced by both microbes and human chemists.

Antibiotic Biosynthesis

Bacteria in the genus *Streptomyces* are the most prolific antibiotic **producers** known, as they produce 60% of all antibiotics in clinical use today. Members of this genus illustrate the virtuosic biosynthetic abilities of bacteria that have changed the course of human history. They are also a perfect example of the potential for future drug discovery from bacteria. Genome sequencing has revealed that many *Streptomyces* species contain dozens of "silent" pathways for antibiotic biosynthesis (Omura et al., 2001). These silent pathways comprise genes that are not expressed under lab conditions. By analyzing the sequences of silent pathways, scientists have been able to predict the compound they would produce if expressed. Many researchers are now trying to unlock these pathways to enable the carriers of these gene clusters to produce novel antibiotics. Some estimates suggest that humans have only discovered 0.1-10% of the natural products of bacteria, which is why **Tiny Earth** is so exciting—this is the treasure trove we aim to discover!

Antibiotic structures and modes of synthesis are linked in interesting ways. Unlike many proteins, which are translated from messenger RNA to produce the final product, antibiotics are not directly encoded in DNA. Antibiotics are produced by complex **biosynthesis pathways**, stepwise chemical reactions catalyzed by enzymes within the cell. They may result from enzyme-catalyzed assembly and modification of amino acids, fatty acids, and sugars. For example, penicillin, a β-lactam antibiotic, is the product of three amino acids – cysteine, valine, and an amino acid intermediate: α-aminoadipate (Figure 7-2) (Clardy et al., 2009). Additionally, some of the enzymes involved in tetracycline (a polyketide antibiotic) and streptomycin (an aminoglycoside) biosynthesis are related to enzymes involved in fatty acid and polysaccharide synthesis, respectively (Clardy et al., 2009; Walsh, 2004). The degree to which these organic substituents are assembled and modified to form an antibiotic helps us group them based on shared chemical characteristics.

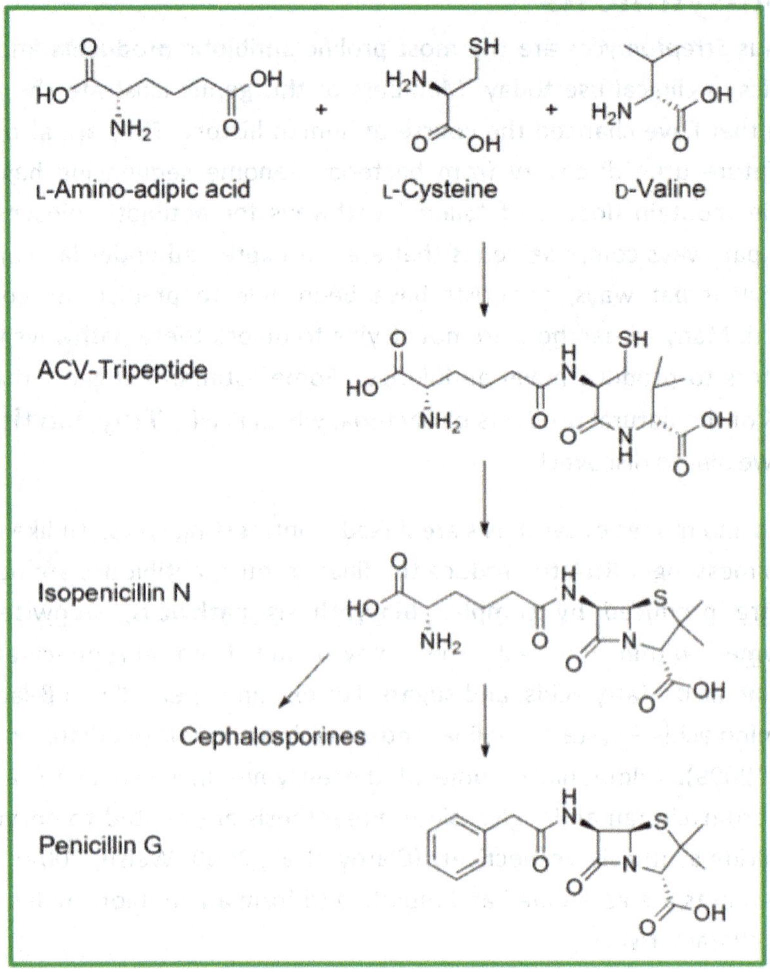

Figure 7-2. Penicillin biosynthesis. Penicillin is the product of three amino acids modified and assembled in an enzyme-catalyzed reaction. Photo source: commons.wikipedia.org

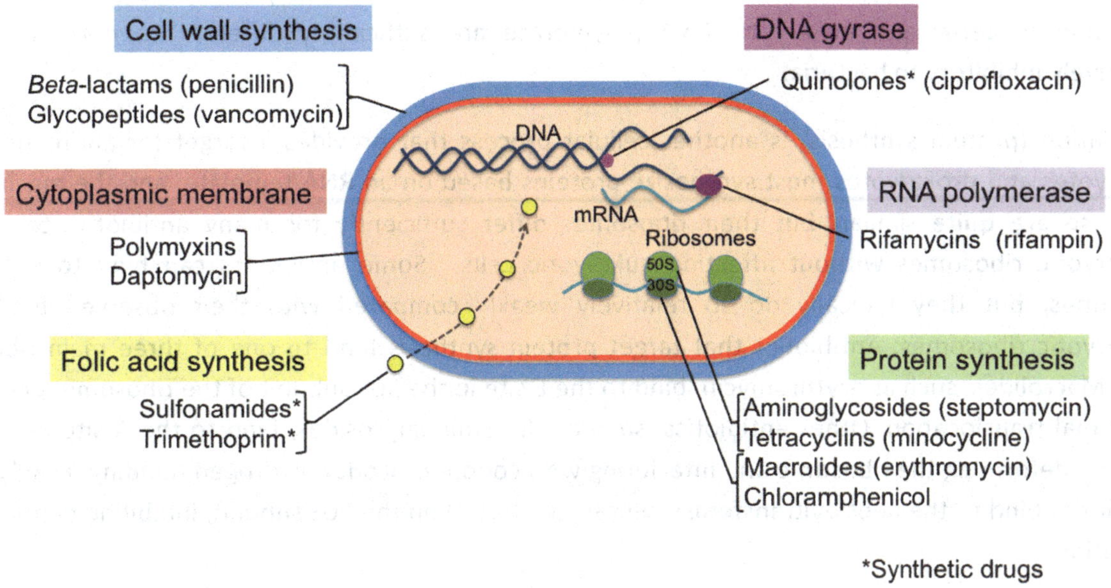

Figure 7-3. Examples of antibiotic cellular targets (modes of action).

Why Antibiotics Kill Bacteria (and not us)

Antibiotics with the same cellular or molecular target are also typically subdivided by their **mechanism or mode of action**. This refers to the specific biochemical interaction that allows the antibiotic to have an inhibitory effect. The fact that antibiotics are selectively toxic to bacteria implies that either they selectively enter bacterial cells or do not recognize a molecular target in eukaryotic cells.

One cellular component unique to bacteria is the cell wall, which is composed of a polysaccharide-peptide (a macromolecule containing amino acids and a complex sugar) called **peptidoglycan**. This tough material protects bacteria from forces in the environment, such as changes in osmotic potential and mechanical force. The wall provides bacteria with structural support and gives them their characteristic shapes. **Gram-positive** bacteria have a thick peptidoglycan layer exposed to the environment, whereas **Gram-negative** bacteria have only a thin layer of peptidoglycan positioned between the cytoplasmic membrane and the outer membrane, which is composed of phospholipids and lipopolysaccharides. Some common antibiotics, such as penicillin and vancomycin, inhibit cell wall synthesis by blocking the enzymes responsible for forming the peptide cross-links between peptidoglycan subunits. Bacterial growth requires constant maintenance and remodeling of the cell wall. When antibiotics interfere with cell wall maintenance, the growing cells will eventually burst (or lyse).

DNA metabolism has properties unique to bacteria. The enzyme DNA gyrase has a similar structure in prokaryotes and eukaryotes, but antibiotics like ciprofloxacin take advantage of small differences in affinity to molecules. DNA gyrase aids in DNA replication by keeping the DNA molecule from overwinding and forming supercoils that block the DNA replication machinery from moving across the template. Similarly, rifampicin inhibits RNA polymerase, which is necessary for transcription of the DNA code and bacterial and eukaryotic RNA polymerase are sufficiently different to make rifampicin selectively inhibitory to bacteria.

Translation (protein synthesis) is another cellular process that provides a target for antibiotics. Both eukaryotes and prokaryotes must synthesize proteins based on an RNA template, and the mechanisms they use are quite similar, but their ribosomes differ sufficiently for many antibiotics to bind to prokaryotic ribosomes without affecting eukaryotic cells. Some antibiotics can bind to eukaryotic ribosomes, but they typically do so relatively weakly compared with their observed binding to prokaryotic ribosomes. Antibiotics that target protein synthesis bind to one of three main ribosomal sites. Macrolides, such as erythromycin, bind to the E site in the 50S subunit of the ribosome, preventing ribosomal translocation. Other antibiotics, such as the aminoglycosides, bind to the A site on the 30S subunit, deforming the ribosome and interfering with codon-anticodon hydrogen bonding. Finally, some antibiotics bind to the peptidyltransferase center (the P site) on the 50S subunit, inhibiting peptide bond formation.

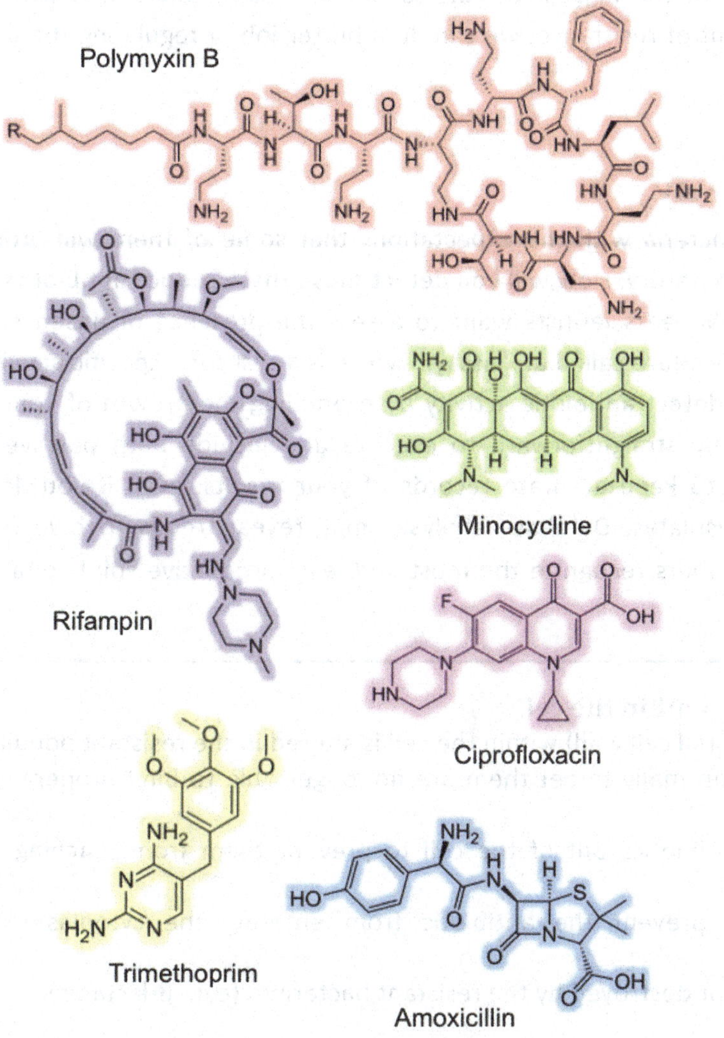

Figure 7-4. Structures of common antibiotics that target various cellular and molecular components. The colors highlighting the chemical structures match the coloring scheme used in Figure 7-3. Red = cytoplasmic membrane, purple = RNA polymerase, yellow = folic acid synthesis, green = protein synthesis, magenta = DNA gyrase, and blue = cell wall synthesis.

Antibiotic Resistance

Antibiotics are effective because they exploit the cellular and molecular differences between bacteria and eukaryotes, targeting necessary functions in bacterial cells, but many bacteria have evolved mechanisms to resist the effects of these compounds. There are two broad mechanisms by which bacteria can acquire resistance to an antibiotic:

1. Bacteria may acquire resistance through a random mutation in a normal gene, or
2. Bacteria may acquire a preexisting resistance gene from another bacterium, most commonly through conjugation.

The rise of antibiotic resistance is one of the biggest threats to human health today, and while we cannot entirely prevent the development of resistance, we can do a better job of regulating the use of antibiotics while developing new drugs.

Antibiotic Production Assay

You have isolated a diversity of soil bacteria with the expectations that some of them will produce antibiotics at detectable levels in the laboratory. How will you detect those that produce antibiotics that inhibit growth of your test organism? When scientists want to assess the presence or absence of a particular component, they design a procedure called an "assay," which is a test for a specific activity of interest. You will conduct an assay to detect antibiotic activity by examining the growth of your test organism. The best assays are relatively straightforward to execute and include both positive and negative controls. It is also important to keep accurate records of your results, even if you do not observe antibiotic activity from your isolates. Database analyses may reveal trends to help future generations of **Tiny Earth** team members recognize the most and least productive soil habitats for antibiotic discovery.

Mechanisms of antibiotic resistance within the cell:

1. Molecular target (e.g., the bacterial cell wall) within the cell is altered in the resistant population so that antibiotics that would normally target them are no longer able to bind properly (e.g., vancomycin).
2. Resistant bacteria transport antibiotics out of the cell to prevent them from reaching their targets (e.g., tetracyclines).
3. The resistant bacterium may prevent the antibiotic from entering the cytoplasm (e.g., lantibiotics).
4. The antibiotic may be modified or destroyed by the resistant bacterium (e.g., β-lactams).

More on antibiotic resistance in Section 10.

References

Amyes, S. G. B. (2001) *Magic Bullets, Lost Horizons: The Rise and Fall of Antibiotics*. New York: Taylor & Francis.

Böttcher, H. H. (1959) *Miracle Drugs*. London: Heinemann.

Brown, K. (2005) *Penicillin Man: Alexander Fleming and the Antibiotic Revolution*. Stroud, Gloucestershire: Sutton.

Clardy, J., Fischbach, M. A., & Currie, C. R. (2009) The natural history of antibiotics. Curr Biol 19:R437–R441. doi: 10.1016/j.cub.2009.04.001.

Nicolaou, K. C., Harrison, S. T., & Chen, J. S. (2009) Discoveries from the abyss: The abyssomicins and their total synthesis. Synthesis (Stuttg) 2009:33–42. doi: 10.1055/s-0028-1083259

Omura, S., Ikeda, H., Ishikawa, J., Hanamoto, A., Takahashi, C., Shinose, M., Takahashi, Y., Horikawa, H., Nakazawa, H., Osonoe, T., Kikuchi, H., Shiba, T., Sakaki, Y., & Hattori, M. (2001) Genome sequence of an industrial microorganism *Streptomyces avermitilis*: Deducing the ability of producing secondary metabolites. Proc Natl Acad Sci U S A 98:12215–12220. doi: 10.1073/pnas.211433198

Walsh, C. T. (2004) Polyketide and nonribosomal peptide antibiotics: Modularity and versatility. Science 303:1805–1810. doi: 10.1126/science.1094318

Other sources:

Butler, M. S., & Buss, A. D. (2006). Natural products — the future scaffolds for novel antibiotics? Biochem Pharmacol 71:919–929. doi: http://dx.doi.org/10.1016/j.bcp.2005.10.012

D'Costa, V. M., Griffiths, E., & Wright, G. D. (2007) Expanding the soil antibiotic resistome: Exploring environmental diversity. Curr Opin Microbiol 10:481-489. doi:10.1016/j.mib.2007.08.009

Harvey, A. (2000). Strategies for discovering drugs from previously unexplored natural products. Drug Discov Today 5:294–300.

Linares, J. F., Gustafsson, I., Baquero, F., & Martinez, J. L. (2006) Antibiotics as intermicrobial signaling agents instead of weapons. Proc Natl Acad Sci U S A 103:19484–19489. doi: 10.1073/pnas.0608949103

Watve, M. G., Tickoo, R., Jog, M. M., & Bhole, B. D. (2001) How many antibiotics are produced by the genus *Streptomyces*? Arch Microbiol 176:386–390. doi: 10.1007/s002030100345

Experiment 7: Design a method to screen for antibiotic producers

Biological Questions:

1. How will you determine whether your soil isolates produce antibiotics?
2. What positive and negative controls will you use in your assays?
3. What factors influence antibiotic production in a microorganism?
4. Can we increase the antibiotic yield of our isolates?

Background:

Objective:

Hypothesis and rationale:

Experimental design and protocols used:

Controls and treatments:

Data and observations:

Interpretations and conclusions:

Notes:

Assignment: Calculating Frequency of Antibiotic Producers

Objective:

Tester organism:

Tester organism media:

Results:

Culture Media	Number of patches tested	Isolates producing antibiotics (number)	Isolates producing antibiotics (%)
	Total:	Total:	

Calculations:

Comments:

Notes:

Section 8: Getting to Know Your Isolates

As we experience the thrill of isolating microbes that produce antibiotic compounds, it is an appropriate time to reflect on our accomplishments thus far and to consider the future direction of our research. Starting with soil samples containing a high density of bacteria, we have isolated microbes capable of inhibiting bacteria related to known pathogens. Furthermore, we have demonstrated that the antibiotic producers are capable of growing in isolation (away from the community of microbes inhabiting the soil) and under laboratory conditions – important traits required for mass production of an antibiotic. But what's next?

We must emphasize that it is a long journey from initial discovery to mass market. There are many criteria that must be met for a new antibiotic to be successful commercially. A few important considerations for clinically useful antibiotics include: Vast amounts of compound are required; toxicity to humans must be minimal at an effective treatment dose; and compounds that remain active after oral ingestion (i.e., in pill form) are more desirable than compounds requiring intravenous administration. A first step in addressing these issues requires isolation of the active compound in pure form. We learned from Fleming's work that this can be a challenge. Separation of the active compound from all of the other compounds produced by the organism requires expensive equipment and often takes months or years of work for even experienced chemists. For this reason, scientists place high value on steps that help to identify, early on, whether the active compound is likely to be new or is a previously identified molecule. We will discuss the purification and structural identification process in Section 9 and Future Directions.

Concurrent with chemical purification and characterization, another approach that lends insight into the potential identity of the unknown molecule is characterization of the producing microbe. For example, *Acremonium chrysogenum* is known to produce cephalosporin C (a β-lactam). If an active soil isolate can be classified as *Acremonium* or a close relative, a priority would be to determine whether the active compound exhibits chemical properties of the cephalosporins. If the antibiotic-producing isolate is a species or genus not shown previously to produce antibiotics, it is of greater interest because the antibiotic has a greater likelihood of being new.

Classifying Microbes

Taxonomy is the field of biology concerned with naming and organizing living organisms into meaningful groups based on similar characteristics. Groups with shared physical features often (but not always) share an evolutionary history. Historically, scientists classified microbes by observing morphological features under the microscope and through biochemical (metabolic) assays that determine the presence of specific enzymes. In the late 20th century, the development of technologies for comparing DNA sequences provided biologists with an approach to classification that also indicates evolutionary relationships and thus gained widespread appeal. Nucleotide sequence similarity provides a way to assess evolutionary relatedness. After all, mutations (heritable changes in the DNA sequence) are the underlying molecular basis for Darwin's "descent with modification." Although whole-genome comparisons are now possible, a bacterial genome may consist of several million base pairs—far too many for quick, routine comparisons. For this reason, scientists typically compare sequences of established "marker genes," those that are present in all microbes and for which function has been conserved. By far, the most commonly used marker gene encodes a small ribosomal subunit (SSU) RNA, the 16S rRNA gene (18S rRNA gene in eukaryotes). 16S rRNA gene comparisons are performed routinely today, but they still do not entirely replace morphological and biochemical analysis. It is important to remember that the strongest taxonomic identifications involve a wide range of techniques, so the results from one method can validate another method. We discuss morphological and molecular taxonomic classification methods in this section and biochemical classification in Section 11.

Molecular Phylogeny

The 16S rRNA gene was first proposed as a tool for comparing microbes by Carl Woese and George Fox in 1977 (Woese and Fox, 1977). At that time, classification of multicellular organisms relied on readily observable morphological features, a system that is not easily applied to microbes because they have far fewer distinguishing anatomical features than biochemical properties. Woese was also interested in evolutionary relationships. Frustrated by the challenges in determining evolutionary relationships among organisms with unimpressive fossil records with few distinguishing visible differences, Woese and Fox searched for a new metric. The idea of using DNA sequences was revolutionary. One feature that makes DNA sequence comparison so valuable is that all living creatures use DNA as their heritable material. Furthermore, all organisms share homologous versions of the SSU genes and these genes evolve very slowly, producing a steady and measurable evolutionary record. For the first time, scientists had a single marker with which to compare the relationships among all living creatures. This resulted in a dramatic rearrangement of what we call the universal tree of life. Comparing 16S and 18S rRNA gene sequences revealed that life is organized in three domains—Archaea, Bacteria, and Eukarya. The discovery that Archaea, although morphologically similar to Bacteria, were as different from Bacteria as are Eukaryotes, along with the prevalence and diversity of microbes revealed by molecular phylogeny, stunned the scientific world. It's sobering to note that all of the Bacteria and Archaea and most of the Eukaryotes (the exceptions are plants, animals, and fungi) are microscopic. We truly live on a microbial planet.

Phylogenetic trees are hypotheses regarding the evolutionary relationships between taxonomic groups; some trees are constructed using SSU sequence data only while others take all available data into consideration. For this reason, trees for the same taxonomic groups may differ (although usually only slightly) depending on the criteria used for constructing the tree. The universal tree of life is based on what we call **molecular phylogeny**, meaning that the relationships presented are based on comparison of molecular features, usually DNA sequence, rather than morphological traits. Increasingly, DNA sequences are used in the construction of trees, but morphological data can still contribute to the complete puzzle when making predictions about evolutionary relationships.

Other genes can serve as markers for comparison, but the 16S rRNA gene remains the gold standard for initial classification. There are many features that make the SSU gene a good yardstick, including its effectively universal presence and convenient size of about 1500 nucleotides; however, its most important advantage over other similar genes is that it contains regions of relatively high variability interspersed among other highly conserved regions. Comparison of the hypervariable regions is useful for identification since these regions typically contain nucleotide sequences that are unique to a particular species.

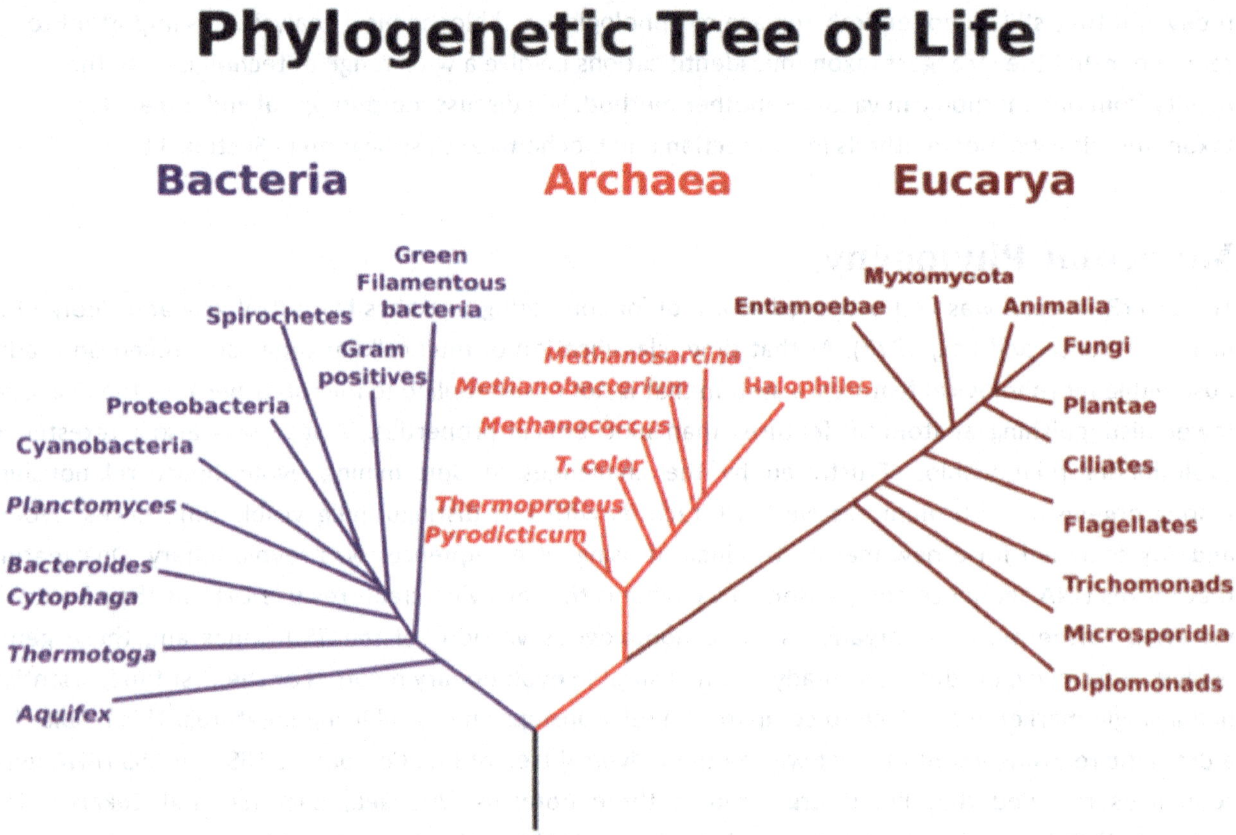

Figure 8-1. The universal tree of life. Photo source: commons.wikimedia.org

Therefore, while SSU sequence comparisons were initially used to create phylogenetic trees for understanding the evolutionary history of life, 16S rRNA gene analysis is now routinely used to develop preliminary hypotheses about the identity of unknown bacterial isolates. This is possible because of the vast collection of known 16S rRNA gene sequences in a publicly available database called GenBank. Search algorithms, such as BLAST, compare an input sequence to millions of known sequences in the database. Once closely matching sequences are found in the database, we can make a preliminary assumption that the unknown falls within the same taxonomic group as the known organisms with highly similar sequences (assuming that the highly similar sequences were themselves reliably identified). More definitive taxonomic assignment can be obtained from observing multiple molecular markers and morphological traits. We will use 16S rRNA gene sequence analysis to make a preliminary identification of our unknown organisms and to make predictions about whether their active molecules are novel.

Obtaining 16S rRNA Gene Sequence Data

DNA sequencing technology has advanced significantly in the past decade; there are now several methods to obtain a DNA sequence for a particular gene or genome. Common to all methods, however, is the requirement for many copies of the DNA destined for sequence analysis (the template). Polymerase chain reaction (PCR) is the technique most commonly employed to amplify (make copies of) the region to be sequenced. We can design PCR primers that are complementary to highly conserved regions of the 16S rRNA gene, thus allowing the same primers to bind to the DNA of a broad range of organisms. Because the amplification products will include the hypervariable regions located between these conserved sites, species-specific sequence data can be obtained for identification and classification purposes.

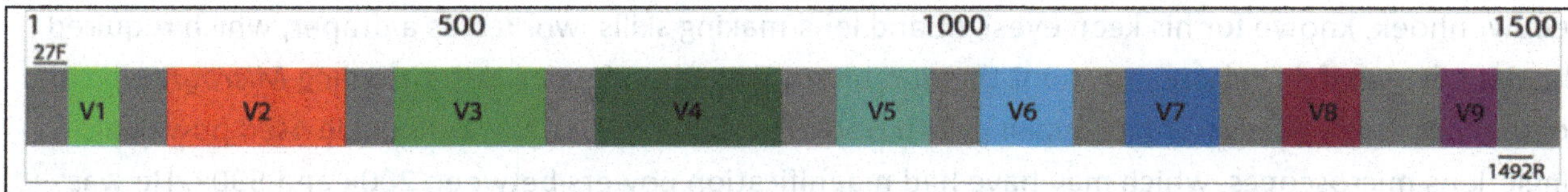

Figure 8-2. 16S rRNA gene illustrating the variable and conserved regions. "V" indicates Variable Regions. Variable regions can be used in group or species-specific applications. Gray boxes are the conserved regions. Conserved regions are used in nonspecific applications. Primers for PCR and sequencing (27F and 1492R) bind to conserved regions to amplify and read all variable regions between the primer set as a template for sequencing.

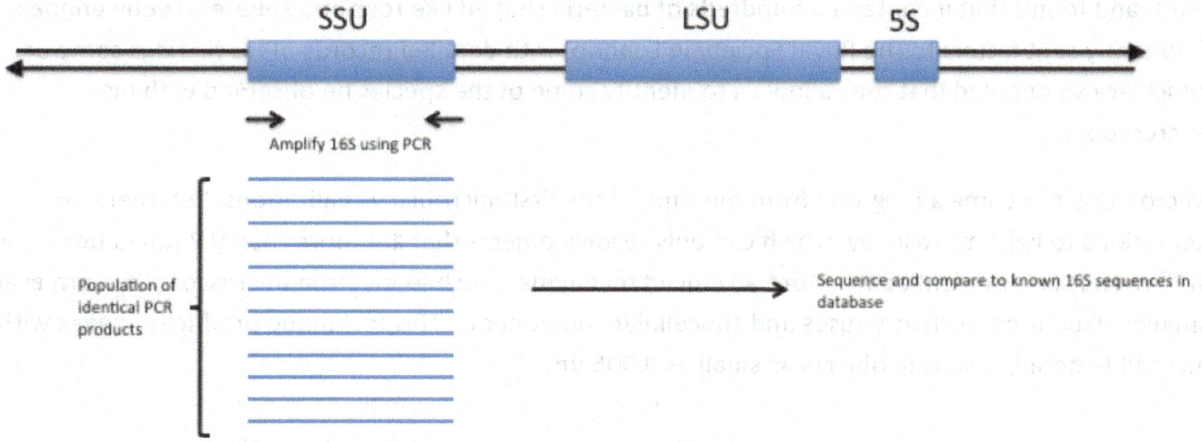

Figure 8-3. Diagrammatic representation of 16S rRNA gene amplification from the prokaryotic chromosome. The small subunit of the bacterial ribosome is encoded by the 16S rRNA gene (SSU), whereas the large subunit is encoded by the 23S rRNA gene (LSU) and the 5S rRNA gene (5S). Schematic adapted from Woese and Fox, 1977.

Morphological Characterization of Isolates

As mentioned, 16S rRNA gene sequence data are exceptionally useful for assigning preliminary taxonomic placement of an organism, but the tentative molecular identification is strengthened through use of several methods. Below, we introduce microscopy as a method to identify key morphological features. Biochemical (metabolic) tests also make up a facet of the identification tools available and will be discussed in Section 11.

Microbes Under the Microscope

The pioneers of microscopy in the late 17th century were also pioneers in the field of biology. Their development of powerful lenses allowed them to delve deeper into the microscopic world of unicellular organisms. Robert Hooke, a renowned Englishman who specialized in the sciences and architecture, is credited with the invention of the compound microscope with which he made detailed observations that are compiled in the book, *Micrographia*, published in 1665, which became a scientific bestseller. The book is famous for its detailed drawings of a fly's eye and the plant cells in a slice of cork. Hooke coined the term "cell" as the cork cell walls reminded him of the cells in monasteries.

Hooke's work inspired a Dutchman who is often called "the father of microbiology," Antonie van Leeuwenhoek, known for his keen eyesight and lens-making skills, worked as a draper, which required the use of magnifying glasses to count the number of threads in fabrics. After reading *Micrographia* and combining his knowledge of lens-making with Hooke's techniques, Leeuwenhoek devised powerful single-lens microscopes, which may have had magnification powers between 200× and 500×. He was very secretive about his techniques and fewer than ten of his microscopes remain. He was such an advanced inventor that it took more than 150 years for scientists to develop microscopes as powerful as Leeuwenhoek's! He was the first to observe and document unicellular organisms, such as bacteria and protists, which he described as "wee animalcules." He famously observed the plaque between his own teeth and found that it contained hundreds of bacteria shaped like rods and spheres. Leeuwenhoek frequently sent letters to the Royal Society in London, with detailed records of his findings some of which are so detailed that they allow us to identify some of the species he observed with his microscopes.

Microscopy has come a long way from the time of the first microbial visualizations. Yet, there are limitations to light microscopy, which can only resolve objects that are more than 0.2 μm (a μm is 1×10^{-6} meter, or a micron) apart. More advanced techniques, such as electron microscopy, discern even smaller structures, such as viruses and subcellular components. This technique produces images with incredible detail, resolving objects as small as 0.005 μm.

Staining the Bacterial Cell Wall

Staining enhances our ability to see the shapes of bacterial cells. Bacteria can be of many different shapes, but the most commonly observed are spherical—called cocci (plural) or coccus (singular)—and rod-shaped—called bacilli (plural) or bacillus (singular). Stains bind to cellular components that would otherwise be transparent under the light microscope. Bacterial cell walls are composed of an organized mesh of carbohydrates, lipids, and proteins. The structural component is a highly cross-linked polymer of sugar and amino acids called peptidoglycan. The peptidoglycan layer protects bacteria and gives them their characteristic shapes. One of the most famous classical staining methods, Gram staining, was developed in the late 1800s to help visualize bacteria in biopsies and has since become crucial for the classification and identification of bacteria by helping distinguish between the two main patterns of bacterial cell wall architecture.

Gram staining employs two stains, crystal violet and safranin, which are purple and pink, respectively. In Gram-positive bacteria, the peptidoglycan layer is very thick (>20 nm, or nanometers) and serves as the outermost wall component. Gram-positive cells retain large amounts of crystal violet in this layer. Once the crystal violet is complexed with iodine, the stains are better able to resist alcohol-acetone decolorization than in Gram-negative bacteria. As a result, Gram-positive cells are left with a purple appearance even after counterstaining with safranin. Conversely, in Gram-negative bacteria, the peptidoglycan layer is thin (normally <10 nm) and covered by an outer membrane. Because the peptidoglycan is so thin, Gram-negative walls are more susceptible to alcohol-acetone decolorization and are rendered colorless when washed with the alcohol-acetone solution. The decolorized cells will thus be left appearing pink once the safranin counterstain has been applied. By discriminating between Gram-positive and Gram-negative cells on a microscope slide, clinical personnel and microbiologists can deduce much information about an isolate's physiology. Furthermore, Gram staining provides an initial broad division that can be used to help classify bacteria. Gram-positives and Gram-negatives make up the two major groups of bacteria, which are generally representative of their evolutionary lineages. Therefore, Gram staining, although a classical method, is still widely used in both clinical and research microbiology to differentiate bacteria and serves as a preliminary method of characterization.

References

Woese, C. R., & Fox, G. E. (1977) Phylogenetic structure of the prokaryotic domain: The primary kingdoms. Proc Natl Acad Sci U S A 74:5088–5090.

Milestones in Microbiology

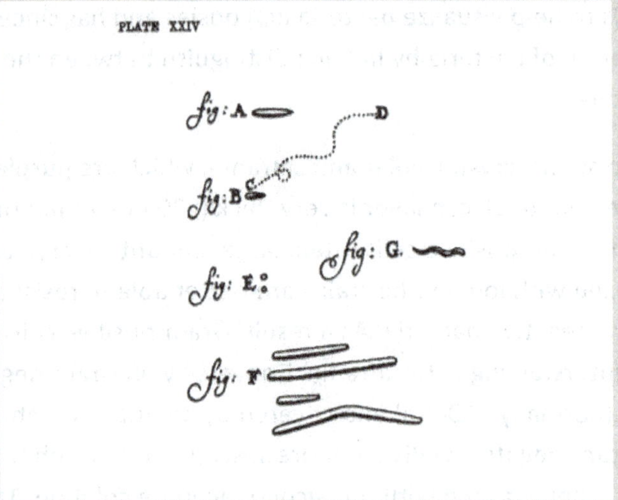

Leeuwenhoek's first representation of the "wee animalcules" he observed under the microscope. This is believed to be the first drawing of a bacterium. Source: University of California Museum of Paleontology – website: http://www.ucmp.berkeley.edu/history/leeuwenhoek.html

Painting of Antonie van Leeuwenhoek by Jan Verkolje. Photo source: https://en.wikipedia.org/wiki/Antonie_van_Leeuwenhoek

Experiment 8: Initial identification of antibiotic-producing isolates

Biological Questions:

1. What do you wish to know about your antibiotic producers?
2. Can we identify bacteria based on their macroscopic morphologies? Why or why not?
3. What cellular and molecular components in our isolates can give us the most information about their identities?
4. How do we confirm what we learn about our isolates?

Background:

Objective:

Results:

Isolate name	**Tentative genus**	**Closest relative (% identity)**	**Size of query sequence (nt)**	**Date of BLAST analysis**	**Gram stain**	**Cell shape**	**Antibiotic production**	**Other**

Based on morphological and/or molecular data, indicate the number of distinct antibiotic producers:

Literature search:

- Morphology: Do your observations match information in the literature?

- Antibiotic production: Do your observations match information in the literature?

Interpretations and conclusions:

Notes:

Section 9: It All Comes Down to Chemistry

Section below adapted from Clayden, J., Greeves, N., Warren, S., Wothers, P. *Organic Chemistry*. Oxford University Press, 2007.

Extraction of Secondary Metabolites or Natural Products

A pure microbial culture produces a vast number of different metabolites, so a first step in the study of a potentially new antibiotic is to separate the active compound from other chemical "contaminants." The purification process usually entails a series of separation steps in which the desirable compound is separated from other molecules in the mix according to its chemical and physical properties. For example, many methods separate compounds based on their relative polarity as indicated by the degree of interaction with various solvents. Contaminants that are chemically similar to the active compound will be the most difficult to separate. Some of the most difficult cases involve enantiomers – molecules that share the same chemical formula and the same chemical bonding pattern, but differ in stereochemistry—they are mirror images of each other. Only one molecule is active; yet it shares most of the same chemical properties as its enantiomeric form, creating difficulties in the purification process.

What Are Organic Molecules?

Organic chemistry started as the chemistry of life, when that was thought to be different from the chemistry in the laboratory. (There was an early theory called "vitalism" that invoked a mysterious force in living organisms.) Organic chemistry then became the chemistry of carbon compounds, and is one of the foundational sciences of biology and the pharmaceutical, perfumery, flavor and dyestuff, and materials industries.

Organic compounds available to us today are those that are present in living things, those produced for over millions of years by petroleum formation, and those synthesized from simple building blocks by synthetic chemists. In earlier times, many of the organic compounds known from nature were those found in "essential oils" that could be distilled from plants or extracted from crushed plant material with acid. Three fine examples are menthol – a flavor extracted from spearmint; *cis*-jasmone – a perfume from jasmine flowers; and quinine – one of the active compounds (against malaria) from the bark of the cinchona tree. Most organic molecules are colorless, but some are known as pigments, which are colored compounds such as indigo, which makes denim blue, and carotene which gives carrots their characteristic color.

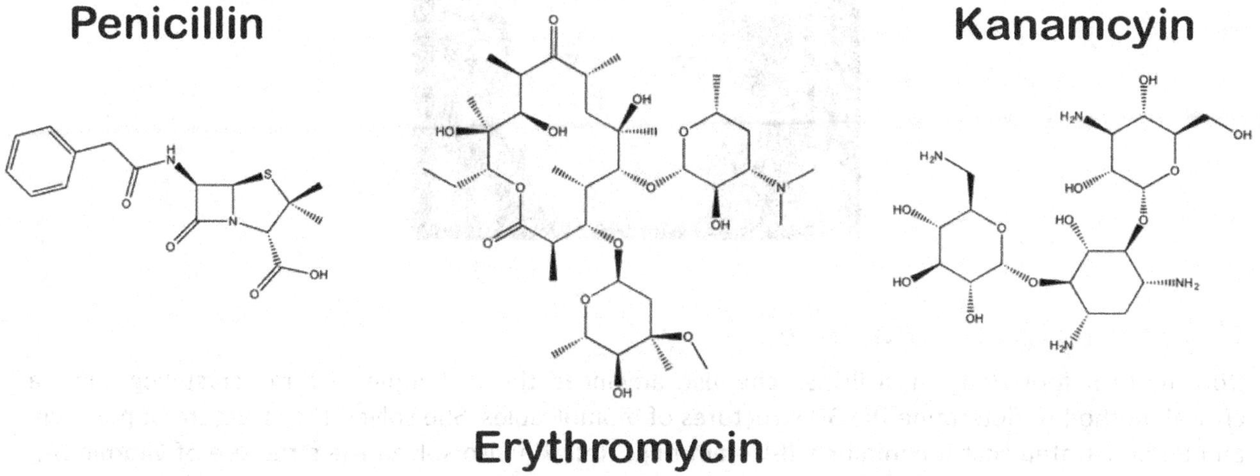

Figure 9-1. [From Left] Chemical structures of penicillin, erythromycin, and kanamycin. Image source: drawn by Nam Kim.

Organic molecules can be crystalline solids, oils, waxes, liquids, or gases. Familiar examples include white crystalline sugar – a natural compound isolated from plants as hard white crystals when pure – and petroleum – a mixture of colorless, volatile, and highly flammable hydrocarbons that include what is commonly known as gasoline, and isooctane, which lends its name to the octane rating of gasoline. In the figure above, isooctane is shown in both a long-hand nomenclature, where all the carbons are shown (lower structure), and the shorthand (upper structure) used by organic chemists to show organic molecules.

What Are Secondary Metabolites?

The macromolecules that are common to all living things (i.e., nucleic acids, proteins, lipids, and carbohydrates) are the products of chemical reactions common to life. These are known as primary metabolic processes. More specialized chemical reactions and molecules are more restricted in their distribution to one or a few species. These molecules are not essential for life, but they are presumed to be important to the organisms in which they are produced, although many have unknown roles. These molecules are often called secondary metabolites, and we have already met quite a few examples in our tour through antibiotics. The majority of known antibiotics are produced as secondary metabolites as the microbial culture reaches the end of exponential growth (refer to the growth curve in Section 5).

Secondary metabolites can range from relatively small molecules that you have already seen, such as penicillin G, up to very large polyethers, including brevetoxin, which is a molecule produced by algae in "red tides" that can kill fish and those who eat the fish.

Milestones in Microbiology

Dorothy Crowfoot Hodgkin, a British chemist, advanced the technique of X-ray crystallography, a critical method to determine the 3D structures of biomolecules. She solved the structure of penicillin and demonstrated that it contains a β-lactam ring. Hodgkin also solved the structure of vitamin B_{12}, which led to the Nobel Prize in Chemistry in 1964. Britain's Royal Society celebrated their 350th anniversary with a set of commemorative stamps, one of which was of Hodgkin and her vitamin B_{12} molecule. Source: http://www.paleophilatelie.eu/description/stamps/uk_2010.html

How Do We Isolate or Extract Secondary Metabolites?

To isolate the compound responsible for antibiotic activity observed in the patch plates, we need to employ separation techniques that are based on characteristics such as size, polarity, solubility, or affinity. There are many useful separation techniques; some are cheap and easy and can be done on a large scale while others require expensive equipment and are time consuming. Each characterization step usually affects the choice of the next one as more is learned about the molecule of interest. For a first step in purification, it is a good idea to try a very general method that has been shown to isolate many secondary metabolites while removing common media compounds and primary metabolites.

End of section adapted from Clayden, J., Greeves, N., Warren, S., Wothers, P. *Organic Chemistry* Oxford University Press, 2007

Solubility of Common Compounds

Dissolves well in water	Do not dissolve well in water
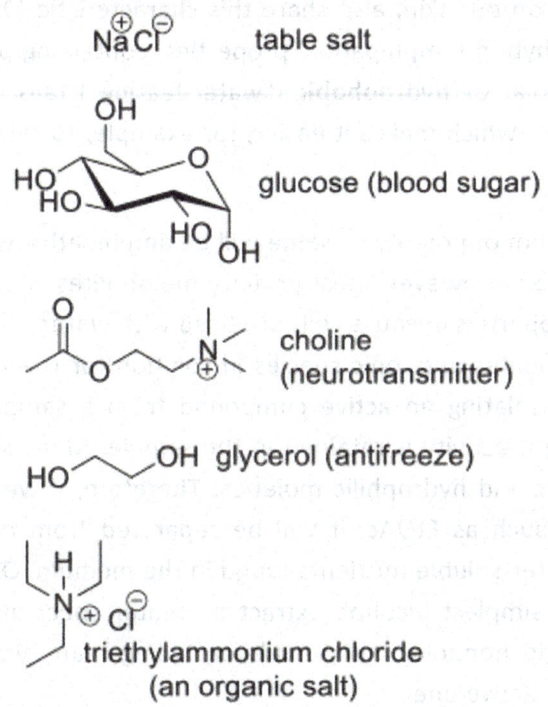 NaCl table salt glucose (blood sugar) choline (neurotransmitter) glycerol (antifreeze) triethylammonium chloride (an organic salt)	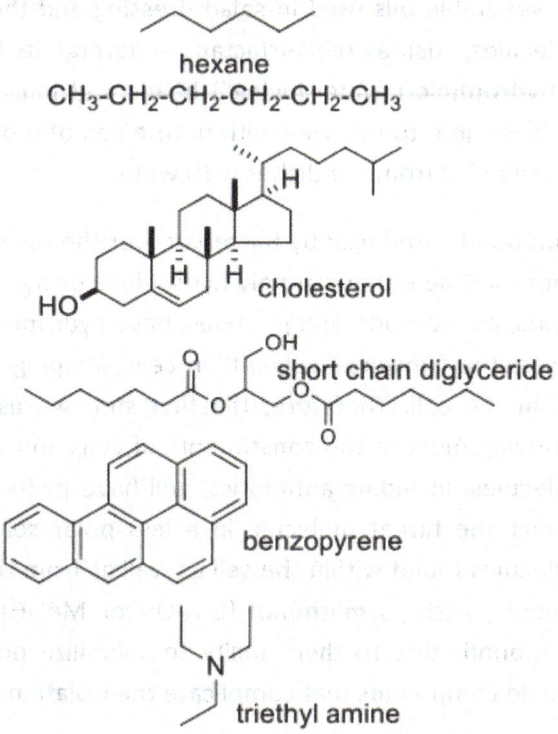 hexane $CH_3\text{-}CH_2\text{-}CH_2\text{-}CH_2\text{-}CH_2\text{-}CH_3$ cholesterol short chain diglyceride benzopyrene triethyl amine

Source: Gillian Phillips.

Extracting with Organic Solvents

While the process of chemical separation to isolate compounds may seem new, similar phenomena surround us. For example, salad dressings demonstrate that oil and vinegar do not mix and form separate layers with different densities and chemical properties. These two liquids also feel very different to our skin. Oil sticks while vinegar and other water-based solutions wash off easily. These familiar interactions are the result of a chemical property known as **polarity**.

Polarity defines the distribution of electron density and charge in a molecule. It results from the tendency of certain atoms to hold onto their shared electrons more closely than other atoms in a bond due to differences in electronegativity. These differences cause some molecules to have "poles" with partially opposing charges; we call these molecules "polar." Water (H_2O) is a great example. Each water molecule has a partial negative charge on the oxygen and partial positive charges on the hydrogen atoms. Atoms in molecules that exhibit this separation of electron density, like other water molecules and water-soluble compounds, will be attracted to other atoms of opposite charges in other polar molecules. These bonds are weak compared with covalent bonds but are strong enough to pull and solubilize polar molecules in a sample. Less polar molecules, such as **ethyl acetate** ($CH_3\text{-}COO\text{-}CH_2\text{-}CH_3$ or

EtOAc) separate from water due to the differences in polarity (similar to oil and water). Even though EtOAc has a "pole" with the carbonyl group (C=O) within its structure, it is not strong enough to associate with water due to the methyl (CH_3) and ethyl (CH_3CH_2) groups. We focus on interaction with water in biology because it is such an important component of biological systems. Triglycerides, such as the vegetable oils used in salad dressing and the oils on our skin, also share this characteristic. Other molecules, such as the surfactant in detergents have hybrid (amphipathic) properties, containing polar or **hydrophilic** ("water-loving") heads and long nonpolar or **hydrophobic** ("water-fearing") tails. This enables them to mix well with both types of molecules, which makes it easier, for example, to remove grease or fat from our dishes with water.

Compounds produced by bacteria cover the full spectrum of polarity — some will be amphipathic while others will be either strongly hydrophilic or hydrophobic. However, most primary metabolites, such as sugars, amino acids, and peptides, have hydrophilic properties because cells are filled with water. Water acts as the "universal solvent" in cells, keeping metabolites and ionic species in solution for reactions within the cell. Therefore, the first step we use in isolating an active compound from a sample is removing most of the constituents of cells and seeing if activity is retained in the sample. Many small molecules, including antibiotics, will have hydrophobic and hydrophilic moieties. Therefore, if we can extract the target molecule in a less polar solvent, such as EtOAc, it will be separated from other molecules found within the cell as well as from the water-soluble nutrients found in the medium. Other solvents, such as **methanol** (CH_3-OH or MeOH), the simplest alcohol, extract a greater spectrum of compounds due to their ability to solubilize polar and nonpolar compounds, including many water-soluble compounds that complicate the isolation of the active one.

EtOAc and MeOH are just two of a wide range of solvents used in extraction protocols as well as in many household products, including glues, nail polish removers, and paint products. Because different solvents solubilize different types of compounds, solubility can suggest the properties of a compound, therefore serving as a useful first step in purification of a compound.

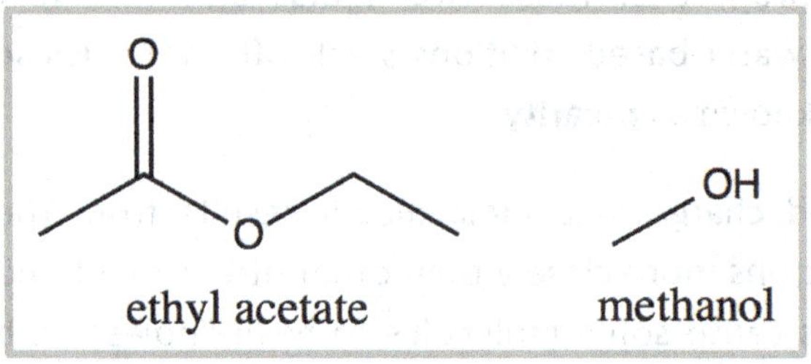

Figure 9-1. Left: Molecular structure of ethyl acetate. This organic compound is nonpolar and serves as a common solvent for other organic, nonpolar compounds. **Right:** Molecular structure of methanol. This compound contains a polar alcohol functional group (-OH), giving it hydrophilic properties. However, its methyl (-CH_3) group, which is slightly hydrophobic, allows it to solubilize both polar and nonpolar compounds to various degrees.

Future Directions

Although the complete purification and analysis of active compounds is likely beyond the scope of your work, it is worth a quick look at the process used to determine the chemical identity of a new molecule.

From Bacterial Isolate to Molecules

Crude extracts such as those created by the extraction process described above are typically complex mixtures of chemicals. This mixture needs to be reduced to a single compound for structural analysis. Therefore, three things are required to isolate to a biologically active molecule through a process called bioassay-guided isolation. It requires a biological assay for activity, purification techniques, and methods to determine the chemical structure.

First, a biological assay is required so the activity can be followed as the purification techniques are used to isolate the active molecule away from other components. Once a pure molecule is obtained, techniques that can identify the molecule due to unique characteristics are used to determine its chemical structure.

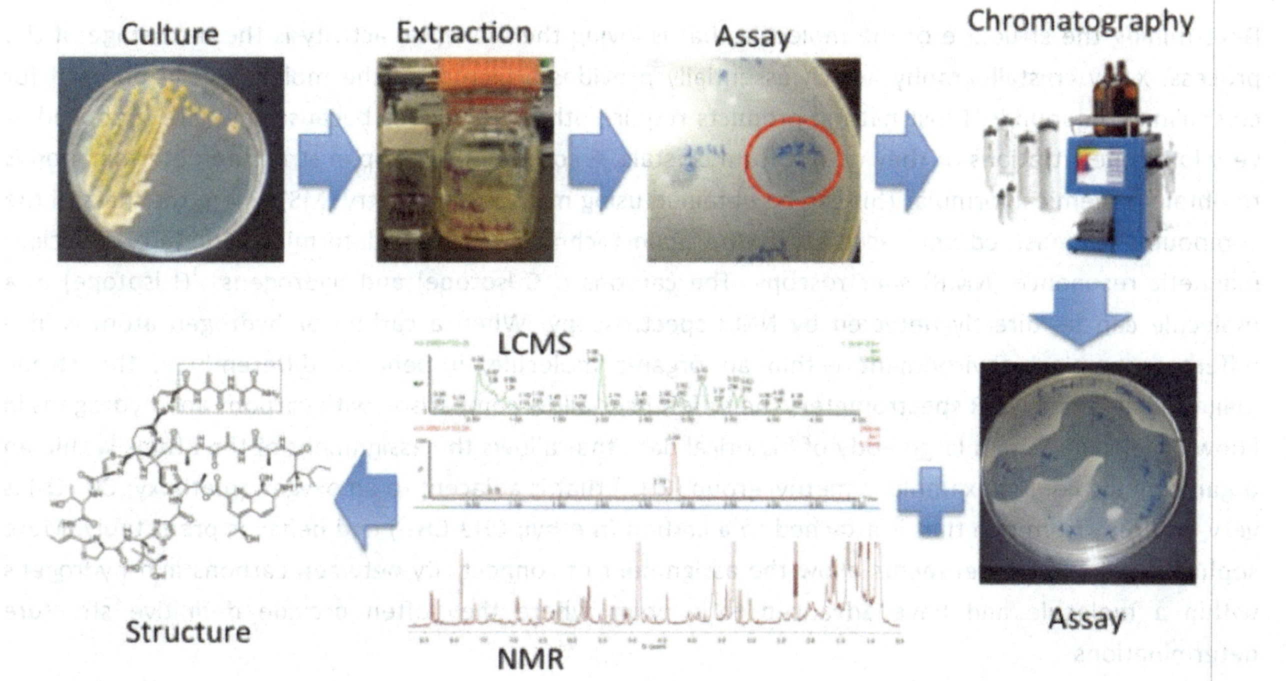

Figure 9-2. Bioassay-guided isolation of thiostrepton from a *Streptomyces* sp. by undergraduate student Kira Garry.

Bioassay-guided isolation is usually time intensive and often leads to known compounds. So before an extract is subjected to the process, it needs to go through a de-replication procedure to determine whether it contains commonly re-isolated compounds, thereby enabling the researcher to prioritize extracts containing novel compounds. Liquid chromatography/mass spectrometry (LC/MS) along with statistical analysis and biological assays that give mode of action information can be used to do this. Once an extract has been prioritized for further work, it will undergo bioassay-guided isolation. A simple example of this is shown in Figure 12-1 where thiostrepton was isolated from a *Streptomyces* sp. by student Kira Garry at Yale University.

1. **Biological Assay**

A biological assay is required before purification starts so the activity can be followed at each step.

2. **Purification Techniques**

The purification of crude extracts is done using various techniques that can be described as liquid chromatography (LC). Highly automated and easy-to-use LC systems are now common in chemistry research laboratories that allow for fast and relatively cheap purification of mixtures. HPLC (high performance LC) systems are also available in cases in which the purification to a single compound is difficult.

3. **Structure Determination Techniques**

Determining the structure of the molecule that is giving the biological activity is the next stage of the process. X-ray crystallography, which essentially provides a picture of the molecule, can be used for crystalline compounds. Most natural products require other techniques because they are produced at very low concentrations or they do not form crystals. A common first step in structure determination is to obtain a chemical formula. This can be obtained using mass spectrometry (MS) where the mass of the compound is measured very accurately. The main technique used to determine structure is nuclear magnetic resonance (NMR) spectroscopy. The carbons (^{13}C isotope) and hydrogens (^{1}H isotope) in a molecule can be directly detected by NMR spectroscopy. When a carbon or hydrogen atom is in a different chemical environment within an organic molecule, it behaves differently in the strong magnetic field of a NMR spectrometer. The differences allow comparison with carbons and hydrogens in known molecules and a large body of historical data that allows the assignment of the atoms within an organic molecule. For example, a methyl group (CH_3^-) that is adjacent to an oxygen (methoxy; CH_3-O-) is very different from one that is attached to a carbon (methyl; CH3-CH_2-) and behaves predictably. More sophisticated NMR experiments allow the assignment of connectivity between carbons and hydrogens within a molecule and have advanced to a point where they often provide definitive structure determinations.

If the compound is known in the literature, this procedure of structure determination can be relatively fast and may take days or weeks. If the compound is new and the structure is similar but not identical to molecules already in the literature, then the process may take weeks to months depending on the complexity of the molecule. In complicated cases, where the carbon backbone has not been observed before, the structure of the novel compound may take months to unravel, particularly if issues of stereochemistry are involved. The mode of action is a key area of interest when novel compounds are isolated. If novel molecules can be identified with new modes of action, this work could have great impact on drug discovery and treatment of infectious disease.

References

Clayden, J., Greeves, N., Warren, S., & Wothers, P. (2007). *Organic Chemistry,* New York: Oxford University Press.

Experiment 9: Test an organic extract of your isolate for antibiotic activity

Biological Questions:

1. Can we isolate an antibiotic from a bacterial isolate? How?
2. What does the choice of solvent for extraction assume about the chemical properties of the antibiotic?
3. How can we assess whether the antibiotic is present in the extract?
4. What else do we need to learn about the observed antibiotic activity?
5. How will you proceed to study your antibiotic extract?

Background:

Objective:

Hypothesis and rationale:

Experimental design and protocols used:

Solvents used:

Solvent pros and cons:

Controls and treatments:

Data and observations:

Interpretations and conclusions:

Future directions:

Notes:

Section 10: Resisting Antibiotics

The antibiotic crisis stems from the increasing prevalence of infectious organisms that fail to respond to treatment with conventional antibiotics, requiring clinicians to prescribe other antibiotics. Occasionally, clinicians have to resort to administering "last resort" antibiotics, which are reserved for the most recalcitrant infections. By the time these antibiotics are deployed, the infection may have spread and made the patient extremely ill and at risk of death. In the worst cases, the infectious organism is resistant to every antibiotic administered. The main reason that the ESKAPE pathogens are such a health threat is not that they, on their own, are particularly virulent; rather, these organisms represent the vast majority of antibiotic-resistant isolates that confound physicians in affected patients. In today's world, it is the acquisition of antibiotic resistance that renders treatment of bacterial infections challenging.

How Does Antibiotic Resistance Arise?

To understand antibiotic resistance, it is helpful to refer back to the mechanism of action of antibiotics and the molecular interactions that occur in cells. An antibiotic is a molecule that enters a cell and interacts with a target molecule; for example, antibiotics that bind to the prokaryotic ribosome have an inhibitory effect on protein synthesis. Any cellular change that prevents the antibiotic from reaching or binding to its target confers resistance to the organism.

There are two main mechanisms through which resistance is acquired. One involves mutations and the other involves acquisition of fragments of DNA harboring genes that confer resistance. Bacteria can acquire DNA from other bacteria, from the environment (released from lysed cells for example), or from viruses through a process called **horizontal gene transfer** (**HGT**). We use the term “horizontal” to distinguish this from the traditional “vertical” transfer of DNA that occurs when a bacterial cell undergoes fission and passes its genes on to a daughter cell. Examples of resistance genes acquired through HGT are genes that code for pumps that expel antibiotics from the cell or enzymes that chemically modify antibiotics thereby inactivating them (i.e., they are no longer reach or bind to their targets).

Mutations that arise through errors in DNA replication can also lead to resistance if: (a) the mutation alters the structure of an antibiotic target such that the binding affinity of the antibiotic is decreased and (b) the function of the target is retained. Remember that antibiotics target structures or functions that are essential for the life of the organism although it may be less effective than the original; only certain mutations meet these criteria.

Antibiotic Resistance in the Environment

When there is selective pressure to maintain the change to the genome, regardless of the mechanism by which the change arose, a resistant population of cells can arise. This can occur in microbes proliferating within patients taking antibiotics to fight infections; from there, resistant bacteria present in feces and bodily fluids can contaminate household or hospital surfaces. Antibiotic-resistant bacteria can spread in the environment through waste disposal, and the use of antibiotics in livestock feed provides further selection pressure for survival of resistant organisms. More than half of antibiotics sold in the United States, for example, are used in livestock production, which has become a topic of contentious debate because most of this use is to promote growth of pigs, cattle, and poultry, not to treat disease (Figure 10-1).

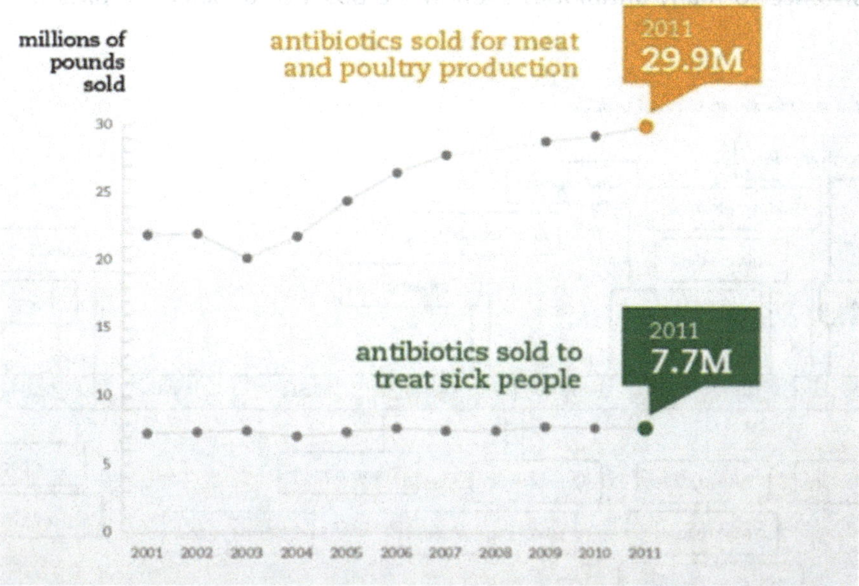

Figure 10-1. U.S. sales of antibiotics, 2001–2011. Photo source: Pew Charitable Trusts http://www.pewhealth.org/other-resource/record-high-antibiotic-sales-for-meat-and-poultry-production-85899449119

Antibiotic use by humans is relatively recent in the history of evolution of life, but antibiotics have been produced by microbes in their natural environments for millennia. The natural role of these compounds is currently unknown, but we can be certain the intended function is not to cure humans of infectious disease. Antibiotics may be one of the most ancient forms of biological warfare. Perhaps secretion of antibiotics is a mechanism employed by microbes to compete with neighboring microbes for limited resources. And as usually occurs in nature, those neighbors that survived the onslaught of antimicrobial compounds proliferated, thus passing on their resistance determinants to subsequent populations of cells either through vertical or horizontal transmission. However, there are only a few examples that have been studied sufficiently to provide strong support for this notion. Alternatively, resistance genes have been speculated to protect microbes from toxins secreted by plants and insects, and antibiotics have been proposed to play a role in cell-cell signaling. The natural role of most antibiotics remains unknown and the subject of active research.

What we do know is that resistant microbes are frequently detected soon after the introduction of an antibiotic for clinical use (Figure 2). Surprisingly, this trend proves true even for compounds created by chemists and synthesized in the lab, for which there has not been selection pressure for resistance in evolutionary history. This highlights the need for more prudent antibiotic use and for the ongoing discovery of new compounds.

In the context of our own research, we might ask whether those microbes that produce antibiotics are likely to harbor antibiotic-resistance determinants. It may seem logical that a microbe would be resistant to an antibiotic that it produces. But how likely is it to be resistant to other antibiotics? And how prevalent is resistance in the environment in the absence of known selective pressure? One important feature in the rapid spread of antibiotic resistance is the fact that antibiotic-resistance genes tend to cluster within the same region of the bacterial chromosome, such that HGT often results in acquisition of resistance to many antibiotics even in the absence of selective pressure for many of the resistance genes.

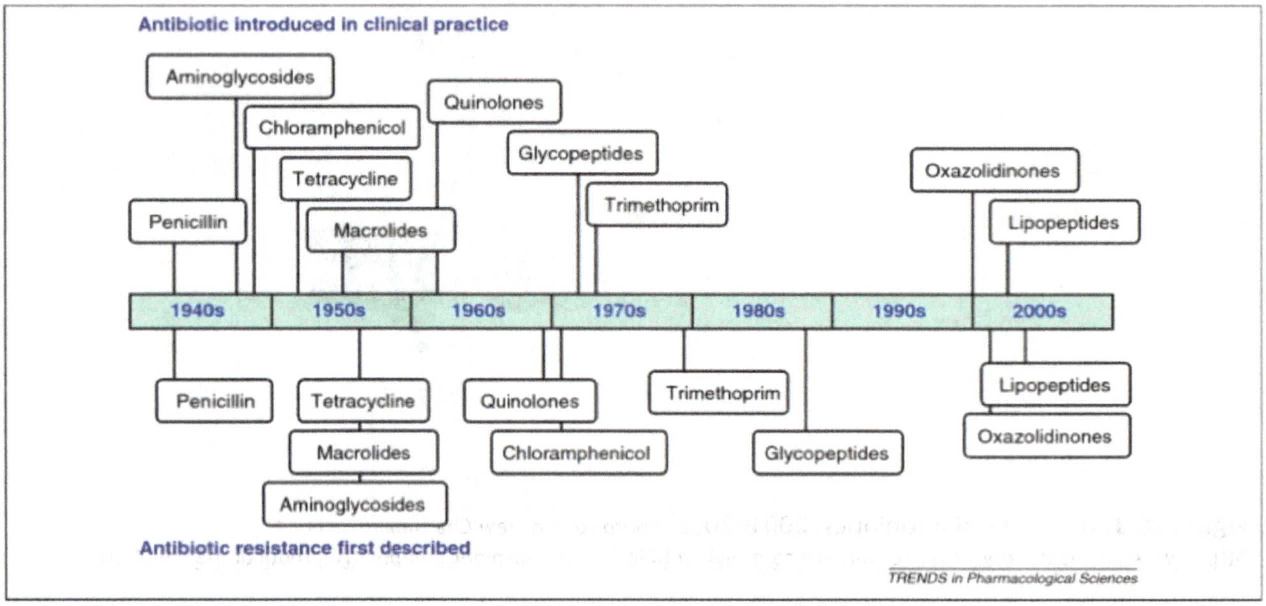

Figure 10-2. Timeline of introduction and appearance of resistance to various antibiotics. Photo source: Högberg, D. L., Heddini, A., Cars, O. (2010) The global need for effective antibiotics: challenges and recent advances. Trends Pharmacol Sci, 31:509–515

Fighting Antibiotic Resistance

The prevalence of antibiotic resistance has called for discovery of new antibiotics and also for the modification of existing antibiotics. At the time of its discovery, penicillin could easily treat a *Staphylococcus aureus* skin infection. As the bacterium acquired resistance, alternatives needed to be developed. In 1959, methicillin was introduced as an alternative to penicillin. Methicillin is a penicillin derivative, which is insensitive to enzymes that would normally break down ß-lactams (*beta*-lactamases). Yet within years of its introduction, resistance was observed. The number of reported cases of methicillin-resistant *S. aureus* (MRSA) has been growing rapidly over the last two decades (Figure 10-3). The last resort antibiotic, vancomycin, has come into common use to treat infections by multidrug-resistant bacteria, and the synthetic antibiotic zylox is now used to treat even more serious infections that are no longer sensitive to vancomycin (Figure 10-4).

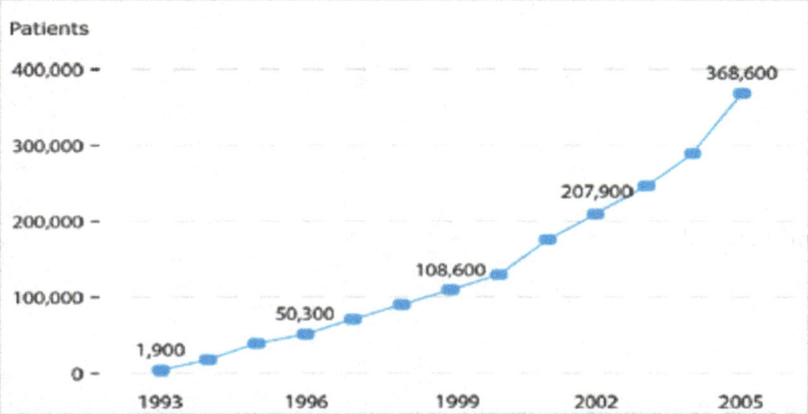

Figure 10-3. Cases of methicillin-resistant *Staphylococcus aureus* (MRSA) annually in the United States. Adapted from "Hospital stays with MRSA infections 1993-2005." Source: AHRQ, Center for Delivery, Organization and Markets, Healthcare Cost and Utilization Project, Nationwide Inpatient Sample, 1993-2005.

References

Allen, H. K., Donato, J., Wang, H. H., Cloud-Hansen, K. A., Davies, J., & Handelsman, J. (2010) Call of the wild: Antibiotic resistance genes in natural environments. Nat Rev Microbiol 8:251–259.

Davies, J., & Davies, D. (2010) Origins and evolution of antibiotic resistance. Microbiol Mol Biol Rev 74:417–433. doi: 10.1128/Mmbr.00016-10

Rice, L. B. (2010) Progress and challenges in implementing the research on ESKAPE pathogens. Infect Control Hosp Epidemiol 31: S7–S10.

Experiment 10: Testing an isolate's resistance to common antibiotics

Biological Questions:

1. Are the antibiotic-producing isolates sensitive to known antibiotics?
2. If so, what does this tell us about the isolates? How might this relate to the antibiotics they produce?

Background:

Objective:

Hypothesis and rationale:

Experimental design and protocols used:

Controls and treatments:

Data and observations:

Isolate ID	**Antibiotic tested**	**Mechanism of action of antibiotic**	**Resistance observed (Y/N)**	**Test organisms inhibited**	**Real world example of bacterial resistance**

Isolates tested	Resistant (%)	Producers tested (number)	Producers resistant (%)

Interpretations and conclusions:

Notes:

Section 11: "Classic" versus "Modern" Classification

Biological Classification

Humans make sense of themselves and the world around them by classifying things based on shared characteristics, whether it is by their physical appearance, function, or origin. For example, classification provides a shorthand for differentiating plants with medicinal value from those that are poisonous, and differentiating animals that walk on their toes from those that walk on their knuckles. Carl Linnaeus, the father of taxonomy, pioneered the classification of organisms based on shared physical characteristics. He meticulously grouped organisms into taxonomic ranks and set the foundation for naming them, using a binomial containing the genus and species designation. Charles Darwin, on the other hand, championed the principles of common descent and paved the way for the classification of organisms based on their evolutionary relatedness. Whether it is looking at morphology or genetic similarity, each system of classification reflects different patterns of evolution and adaptation to environments, distribution in time and space, or utility to humans.

Whatever classification system we employ, it is important to know the criteria used to differentiate things. The use of molecular phylogeny has become the standard for classifying bacteria based on their genetic relatedness, but this is usually supported by morphological and biochemical characterization techniques as discussed in Section 8. To classify and identify bacteria, microbiologists assess many features such as the structure or function of key biomolecules; the shape, size, or staining pattern of the cells; the type of metabolism; or DNA sequence composition.

In this section, we will describe some of the most common biochemical characterization methods used in clinical diagnostic labs to identify bacterial pathogens isolated from patients. They are also applied in research labs to establish taxonomic and phylogenetic relationships, providing a better understanding of shared characteristics, evolution, and adaptation to natural environments than could be obtained with any single method.

The Diagnostics Lab

When it comes to diagnosing and treating patients, pathogen identification must be fast, reliable, and low cost, whenever possible. Classical methods of identification are based on fundamental microbiological principles about how bacteria respond to certain changes to their environment. For example, changing the chemical or nutrient composition of a growth medium affects how bacteria grow or what chemicals they produce. To exploit these changes, two types of specialized media are used — **selective media**, which support the growth of some species and not others, and **differential media**, which differentiate species by visible characteristics.

In hospitals, the use of selective media is a widespread method of identifying human pathogens. Bacteria may be isolated from bodily fluids (e.g., urine or blood), abscesses, or other areas of infection in the patient's body. These samples are streaked out on specialized media that accomplish different objectives. To identify methicillin-resistant bacteria present in a sample, we can formulate a selective medium containing methicillin, which will inhibit the growth of all susceptible bacteria but not the resistant ones – a simple process of elimination. Selective media may also lack an essential amino acid, which would inhibit bacteria unable to synthesize that nutrient. Differential media, on the other hand, will help distinguish groups of bacteria through changes in appearance of the organism or the medium itself, usually in response to differences in biochemical pathways. Researchers and clinical personnel use these visual cues to assign phenotypes and chemical reactions associated with known bacteria, pathogenic or not.

Specialized Media and Enzymology

As described in Section 8, Gram staining allows us to assign bacteria to one of two large groups of classification: Gram-negative or Gram-positive. Other tests do this while conveying even more information about the microbe's biochemistry. MacConkey agar is a specialized medium that differentiates between Gram-negative bacteria, especially those that colonize the human gastrointestinal (GI) tract. This medium inhibits the growth of Gram-positive bacteria by disrupting their cell walls with crystal violet and bile salts infused in the medium. Many Gram-negative bacteria, such as *Escherichia coli*, thrive in the GI tract as part of our natural gut flora along with many specialized Gram-positive bacteria. Other Gram-negative bacteria, such as *Salmonella* and *Shigella*, are notorious pathogens that can cause food poisoning and even death. To differentiate between these bacteria, MacConkey agar allows us to deduce information about their ability to ferment carbohydrates. A pH indicator in the medium changes the appearance of cells producing acid, a by-product of fermentation, from their original color to pink/red. *E. coli* ferments the lactose present in the medium and releases acid, making colonies and the surrounding agar appear red in the presence of a pH indicator. Conversely, *Salmonella* and *Shigella* do not ferment lactose and maintain a neutral pH, making colonies display their typical white or tan color (Wessner, et al., p. 176). MacConkey agar is therefore both selective and differential, revealing information about the cell wall composition and ability to ferment lactose.

Another method to differentiate enteric bacteria based on their ability to metabolize different compounds is the triple sugar iron (TSI) test. TSI medium, which consists of solidified agar in a slanted test tube, supports classification of bacteria based on their ability to ferment different carbohydrates (glucose, sucrose, and lactose), to produce H_2S (hydrogen sulfide, a by-product of bacterial anaerobic digestion), and to grow in the presence or absence of oxygen (ASM Microbe Library). In this method, bacteria are stabbed into the agar, where they begin to metabolize sugars under aerobic conditions at the surface of the agar or anaerobic conditions at the butt of the tube. A pH indicator turns the medium from red to yellow in response to acid as a by-product of fermentation. For example, a lactose fermenter like *Escherichia* or *Klebsiella* will continue to lower the pH long after glucose and sucrose have been depleted in the medium. Since they are both facultative anaerobes, they will ferment sugar both in the presence and absence of oxygen, making the whole tube appear yellow. In contrast, *Citrobacter* and *Salmonella* will only ferment carbohydrates under anaerobic conditions and release H_2S. Therefore, only the butt of the tube will appear yellow, and black precipitate will form from the reaction of H_2S with ferric ions in the medium (Wessner et al., p. 175).

Similarly, the starch agar test differentiates between bacteria that hydrolyze starch, a branched polymer of glucose units, and bacteria that do not. For this method, bacteria are grown on starch agar. When colonies become visible, the plate is flooded with iodine, which is normally red but turns blue when it complexes with starch, serving as an effective indicator for the presence or absence of starch molecules. A starch hydrolyzer like *Bacillus subtilis* will secrete amylase, which breaks down the surrounding starch to glucose. Upon flooding the colonies, the medium surrounding the colonies appears red in the absence of starch, which has been broken down by the amylase-positive bacteria. The rest of the plate will appear a dark shade of blue where the starch is still present. *Streptococcus* and *Staphylococcus* are both starch negative and can be used as controls.

By testing a bacterium's ability to metabolize different compounds, we are indirectly screening for the presence of key enzymes that are unique to certain taxonomic groups of bacteria. Bacteria are equipped to survive in distinct environments where different carbon and energy sources are available. They are also equipped to deal with a variety of mechanical, chemical, and environmental conditions that threaten their existence. For example, hydrogen peroxide (H_2O_2) is a strong oxidizing agent that rapidly kills susceptible cells and is employed by many organisms to protect against infectious bacteria. H_2O_2 is also used as an antiseptic applied on cuts or wounds on the skin. Nevertheless, many bacteria have evolved a mechanism to counteract H_2O_2 with the enzyme catalase. *Staphylococcus* bacteria, like those that colonize our skin, and *Bacillus*, which is commonly found in the soil, both test positive for the presence of this enzyme. Testing positive for catalase is made evident by exposing target bacteria to a 3% H_2O_2 solution and observing for effervescence. When catalase acts upon H_2O_2, it produces water and oxygen gas.

$$2H_2O_2 + \text{catalase} \rightarrow 2H_2O + O_2$$

When dealing with common or well-studied bacteria, these classical biochemistry-based tests can give us great information about our bacterium to narrow down its possible identity. Diagnostic labs with well-developed identification keys can use these methods to type pathogens down to one or several genera and figure out a line of treatment for patients. However, to identify a bacterium down to the species level, or when dealing with a poorly studied bacterium, sequencing technology is normally necessary.

Classic or Modern Approach

Over the past two decades, sequencing technologies have become faster, more reliable, and cheaper, leading to the exponential increase of sequenced genomes available in online repositories like GenBank and other databases. 16S rRNA gene sequencing has therefore become a useful method for identifying bacteria down to the genus level, and more specific primers can help us identify their species or particular strain. Knowing that a bacterium's 16S rRNA gene sequence has a 97% or higher identity to a known species provides us information about an organism of interest and what phenotypes to expect. As we indicated in Section 8, researchers usually combine morphological, biochemical, and molecular analyses to provide a complete picture of the identity of an unknown organism. Results may bring to light discrepancies that may indicate we are dealing with a new strain or a variant of a known species. Whatever the case, coupling knowledge of a bacterium's phylogenetic background with morphology and biochemical information acquired through classical biochemistry-based methods allows us to draw a comprehensive picture of our isolates and acquire new insight into their cellular and molecular processes.

References

ASM Microbe Library. *Laboratory Protocols*. <http://microbelibrary.org/about/51> Date accessed: 3 September 2013.

Fox, A. Culture and Identification of Infectious Agents. Bacteriol Microbiol Immunol, On-Line: University of South Carolina School of Medicine <http://pathmicro.med.sc.edu/fox/culture.htm> Date accessed: 3 September 2013.

Wessner, D. R., Dupont, C., & Charles, T. (2013) *Microbiology* New York: Wiley & Sons.

Experiment 11: Biochemical characterization of isolates

Biological Questions:

1. Now that you know the genus of your isolate, what else do you want to know about it?
2. Is your biochemical characterization information consistent with previous characterization of this genus?

- Note: Reproduce the following prompts for every biochemical test done. -

Background:

Objective:

Biochemical test:

Procedure:

Results:

Interpretation:

Is this consistent with what you know about your isolate(s)?

Conclusions:

Notes:

Section 12: Bacteria in Context

"Good" versus "Bad" Bacteria

The popular belief that all bacteria are "germs," which the Merriam-Webster's Dictionary defines as "microorganisms causing disease," has led to great efforts to eradicate bacteria from our lives with stringent sanitary and hygienic measures. Conversely, the food industry has marketed "good bacteria," or probiotics in fermented products such as the "live and active culture" label on yogurts. This dichotomy has led to the notion that, in general, bacteria cannot be trusted, but can have beneficial effects if consumed from the right packages. However, the notion of discretely "good" and "bad" bacteria is misleading. The truth is that most bacteria are harmless and essential components of the Earth's ecosystems and host organisms. From birth, bacteria colonize our skin and guts, enhance our immune systems, and enrich our diets with nutrients that would otherwise be inaccessible to us as humans. Recent research indicates that gut bacteria play roles in protecting us from diseases as diverse as autism, asthma, diabetes, depression, and obesity. The interdependence between bacteria and other organisms is not coincidental, but is the product of a long history of coevolution and interspecies interactions.

Symbiotic Relationships

Relationships define the way we, humans, interact with each other. They define how we communicate, the way we feel toward one another, the agreements and commitment we make, and the norms that dictate those interactions. They can last a short time or a lifetime, and they often lay a foundation for those same relationships in future generations. As social beings, and as living things, we are fundamentally interconnected and interdependent.

Symbiosis is a long-term, intimate relationship between two or more species (Wessner et al., p. 563). Like all relationships, symbioses may serve various purposes for each of its participating members, whether it is receiving essential nutrients, providing shelter and protection, or disseminating progeny. These relationships often fall in a continuum between beneficial and harmful for one species versus the other. On one end of the spectrum, a **mutualistic** relationship defines an interaction in which both species benefit. For example, herbivores rely on their gut bacteria to break down cellulose, and the gut bacteria in return are provided a habitat and a steady food supply. At the other end of the spectrum, a **parasitic** relationship defines an interaction where one species benefits and the other is harmed. Bacteria that have observable or measureable effects on their hosts may be mutualistic, parasitic, or fall anywhere in between.

Residing On or Inside Another Organism

Symbioses are not only defined by their effects (beneficial versus harmful) on the host, but also by their location in relationship to the host. The species involved in a symbiotic relationship are known as **symbionts**. **Ectosymbionts** reside on the surface of their hosts, while **endosymbionts** reside in the tissues or within the cells of their host (Wessner et al., p. 563).

Most scientists now agree that eukaryotic cells are the product of a form of endosymbiosis. When biologist Lynn Margulis published a paper supporting the **endosymbiotic theory**, many scientists were quick to reject the idea that mitochondria and chloroplasts were once "free-living prokaryotic cells" engulfed by early eukaryotic cells (Sagan, 1967). Today, molecular and genetic evidence supported by 16S rRNA sequencing of mitochondrial and chloroplast DNA indicates that these organelles may have originated from α-Proteobacteria and cyanobacteria, respectively. The process of symbiosis was perhaps the single most important event in the evolution of complex organisms.

Figure 12-1. A mealybug feeding on sap. Photo source: http://schaechter.asmblog.org/schaechter/2011/09/a-bug-in-a-bug-in-a-bug.html

Figure 12-2. A leaf-cutting worker ant covered by the ectosymbiont *Streptomyces*. Leaf-cutting ants rely on the antibiotics produced by *Streptomyces* species to keep their "fungus farms" parasite-free. Photo source: http://scienceblogs.com/notrocketscience/leafcutter-ants-rely-on-bacteria-to-fertilise-their-fungus-g/ (accessed 17 July 2014)

In the long term, some symbiotic relationships irreversibly alter host physiology, biochemistry, and genetics. The human genome with 21,000 protein-coding genes and the *E. coli* genome with just over 4,000 are both enormous in comparison with certain bacteria that rely on other members of a three-way symbiosis. The bacterium *Candidatus Tremblaya princeps* contains only 120 protein-coding genes, an unprecedented low number for any living organism. *Tremblaya*, an endosymbiont of the mealybug, synthesizes vitamins and amino acids from the sap consumed by its host in a mutually beneficial relationship. Yet, *Tremblaya* is itself dependent on its own endosymbiont, the γ-Proteobacterium *Moranella*, whose genome is significantly larger, to synthesize proteins that are not encoded in *Tremblayas*'s minimalist genome (Husnik et al., 2013). The interdependence displayed in this three-way symbiosis has meticulously divided labor among these organisms, all in the safety and steady food supply of the mealybug host (Figure 12-1).

Streptomyces, which have some of the largest bacterial genomes and produce most commercially and clinically used antibiotics, are also involved in symbiotic relationships thanks to their bioactivity against pathogens. Leaf-cutting ants rely on antibiotic producers like *Streptomyces* to protect their food from microbial infection (Scott et al., 2008). These so-called farming ants cultivate the fungus *Leucoagaricus* for food, which is prone to infection by yet another fungus called *Escovopsis* (Currie et al., 1999; Seipke et al., 2011; Haeder, Wirth, Herz, & Spiteller, 2009). *Streptomyces* lives on the ants' exoskeletons as an ectosymbiont and produce antibiotics that specifically inhibit *Escovopsis* (Seipke et al., 2011). This protects the ants' cultivar fungus as their source of sustenance and provides shelter for the antibiotic producers living on them (Figure 12-2).

Plants and Microbes

Anywhere you look on the planet, plants interact with microbes in the soil, on their surfaces, and even living within their tissues. A few of these interactions are parasitic and are the cause for great loss in agriculture. Like the *Escovopsis* fungus threatening the crops of leaf-cutting ants, species of the eukaryotic microorganism *Pythium* attack the root tips of plants. Almost all plants are susceptible to *Pythium*, which, under the right conditions and with no competition, can spread very quickly and ravage fields with root rot (Moorman, 2013). The use of biological control agents, such as bacteria that enhance crop growth or inhibit plant disease, has been purposely applied in agriculture to restore the balance. Thus, our dependence on bacteria for our food source is tightly intertwined with the practices of modern agriculture.

Bacteria have been used for millennia to enhance crop nutrition. Nitrogen gas (N_2) constitutes roughly 80% of the Earth's atmosphere. Nitrogen is an essential component of amino acids and nucleotides, but N_2 gas is inaccessible to the vast majority of organisms because of the ultra-stable triple bond between the N atoms. Genes encoding nitrogenase, the enzyme that converts N_2 from the atmosphere to NH_4^+ (ammonium), have only been found in Bacteria and Archaea (Wessner et al., p. 474). The conversion of nitrogen gas into ammonium is known as **nitrogen fixation**, an essential part of the nitrogen cycle. Ammonium and its deprotonated form, ammonia (NH_3), can then be assimilated by many organisms, like plants, into essential biological molecules (Figure 12-3).

Leguminous plants (e.g., beans, peanuts, and peas) are protein rich and therefore have a high demand for nitrogen, which is an abundant element in protein. To obtain their nitrogen, legumes have evolved living quarters for nitrogen-fixing bacteria. Bacteria that live in the soil with a plant host and fix nitrogen are collectively known as **rhizobia**. In response to contact with rhizobia, legumes form **root nodules**, tumor-like organs that house the rhizobia in a perfect partnership. The plant secretes a chemical signal that invites *Rhizobium* to its root hairs (elongated cells on the surface of the root), and once the microbe-plant contact occurs, the root hair curls around the bacteria, enveloping them inside the root cell (Needham, 2000). The rhizobia spread into the root cortex through a plant structure called an infection thread, and in the cortex, the bacteria lose their cell walls as they make their home within plant cells and become nitrogen-fixing bacteroids. The root nodule provides an anaerobic environment for nitrogenase activity. In order to provide oxygen (O_2) for other reactions without interfering with nitrogen fixation, plants produce the heme protein leghemoglobin, which carries oxygen to the cells while simultaneously sequestering it away from nitrogenase (Wessner et al., p. 566). This interdependence between legumes and rhizobia demonstrates a shared evolutionary history.

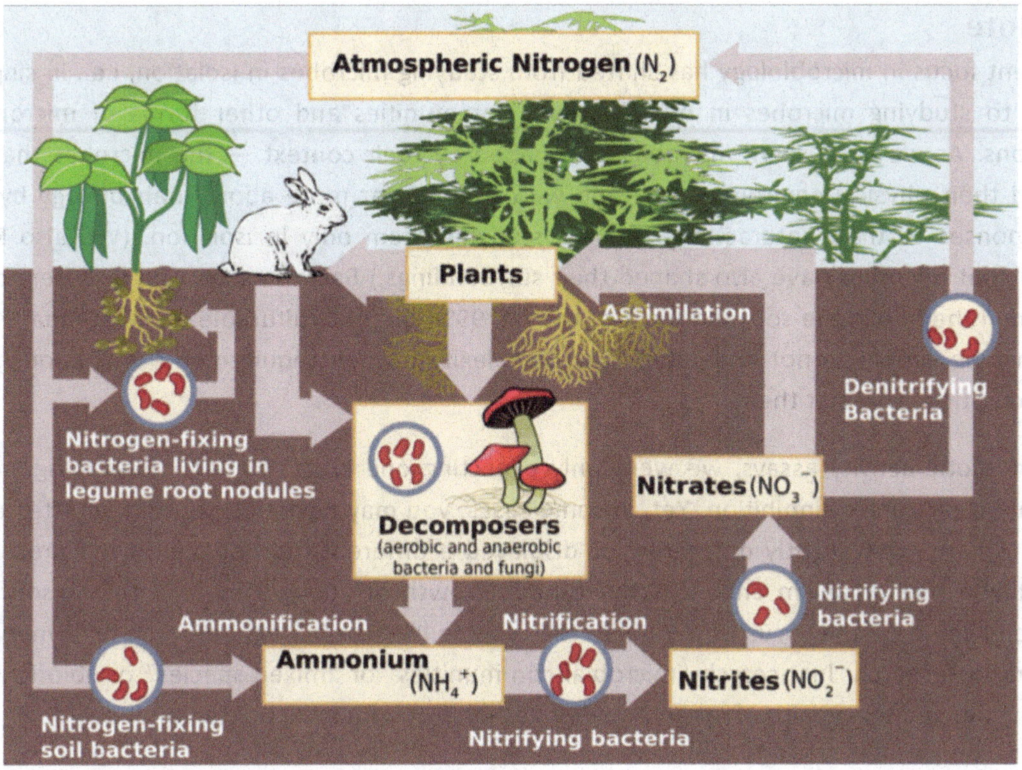

Figure 12-3. Schematic of nitrogen cycle. Bacteria play an essential role in the nitrogen cycle by "fixing" nitrogen (converting N_2 gas to ammonia) and making it usable for plants. Photo source: commons.wikimedia.org

Microbes and Us

Microbes not only make up most of the cells and DNA inside and on our bodies, but also play important roles in our digestive system, immunity, and probably even behavior. Whether we acquire them from things we touch, from yogurt or other fermented products, or from each other or other species, categorizing them as either "good" or "bad" is not straightforward. The Human Microbiome Project, an initiative launched by the U.S. National Institutes of Health, seeks to characterize all the entities that make up the **human microbiome**, or collection of microorganisms in our bodies. Articles challenging traditional views of microbes and urging us to "tend" our microbial flora with "good" bacteria-promoting diets and even microbiota transplants (Zimmer, 2012) permeate the media. Nevertheless, bacteria that can enrich our nutrition and prevent pathogenic bacteria from colonizing the gut epithelium can be closely related to the bacteria that cause food poisoning and lead to massive product recalls. That is because bacteria, like all organisms, exist in their current state in context. Microbes in a **consortium**, or a collection of symbiotic species in a microbial community, are kept in check by neighboring organisms but display different phenotypes in isolation. This creates a challenge when studying microbes outside of their context, whether that is a set of abiotic conditions or the inside of mealybug.

Final Note

The current focus in microbiology has shifted from studying microbes in isolation, i.e., in single-species cultures, to studying microbes in mixed-species communities and other forms of microbe-microbe associations. As we have seen, microbes are shaped by their context — the microbes that surround them and their physical environment — and we can learn far more about their biology by observing their responses to these interactions than by studying them only in isolation. (We also know from Section 1 that microbes have also shaped their surroundings.) Bacteria grow and look very different in the lab than they do in the soil, and most of them (~99%) are not culturable under standard laboratory conditions; we simply do not understand the complexity of their requirements well enough to design culture conditions that suit them.

In our antibiotic activity assays, we were only expecting one type of interaction: antagonism, which results in a clear zone of inhibition. Yet, in some cases, you may have noticed that your tester strain or bacterial isolate grew slightly differently or displayed a different morphology in the presence of the other. Maybe one organism enhances the other's growth or cooperates to access resources. These interactions could be the focus of a whole other course in microbiology research or someone's entire life's work in research. Just search "microbial communities" or "mixed-species" or "biofilms" on your search engine, and the wealth of up-and-coming research you find will be tremendous!

It is important to recognize that our approach to studying microbes has been very targeted, and that just as microbes are abundant and diverse, so are the possibilities of research. As you go on to pursue the chemical structure of your antibiotic or study its spectrum and mechanism of action in more detail, keep in mind that each bacterium contains a wealth of information, which grows as you study it in the context of other organisms or other variables. You have already gained a strong foundation to talk about your microbes, learn more about them in the literature, and design some final experiments. The insights you have gathered about them are the product of your interactions with the organism, and in many ways, *you* have become their context. This is the nature of research. Our experiences leave an indelible mark in the world, and with this research, you have left yours.

References

Currie, C.R., Scott, J.A., Summerbell, R.C., Malloch, D. (1999) Fungus-growing ants use antibiotic-producing bacteria to control garden parasites. Nature 398:701-704.

Haeder, S., Wirth, R., Herz, H., & Spiteller, D. (2009) Candicidin-producing *Streptomyces* support leaf-cutting ants to protect their fungus garden against the pathogenic fungus *Escovopsis*. Proc Natl Acad Sci U S A 106:4742–4746.

Husnik, F., Nikoh, N., Koga, R., Ross, L., Duncan, R. P., Fujie, M., Tanaka, M., Satoh, N., Bachtrog, D., Wilson, A. C., von Dohlen, D. D., Fukatsu, T., & McCutcheon, J. P. (2013) Horizontal gene transfer from diverse bacteria to an insect genome enables a tripartite nested mealybug symbiosis. Cell 153:1567–1578. doi: 10.1016/j.cell.2013.05.040

Moorman, G. W. (2013) Plants & Pests: *Pythium*. Penn State College of Agricultural Sciences. http://extension.psu.edu/pests/plant-diseases/all-fact-sheets/pythium

Needham, C. (2000) *Intimate Strangers : Unseen Life on Earth*. Washington, DC: ASM Press.

Sagan, L. (1967). On the origin of mitosing cells. J Theor Biol 14:255–274.

Scott, J.J., Oh, D.C., Yuceer, M.C., Klepzig, K.D., Clardy, J., Currie, C.R. (2008) Bacterial protection of beetle-fungus mutualism. Science 322:63.

Seipke, R. F., Barke, J., Brearley, C., Hill, L., Yu, D. W., Goss, R. J. M., & Hutchings, M. I. (2011). A single *Streptomyces* symbiont makes multiple antifungals to support the fungus farming ant *Acromyrmex octospinosus*. PLoS ONE 6(8):e22028. doi: 10.1371/journal.pone.0022028

Wessner, D. R., Dupont, C., & Charles, T. (2013) *Microbiology* New York: Wiley & Sons.

Zimmer, C. Tending the Body's Microbial Garden. The New York Times. June 18, 2012. http://www.nytimes.com/2012/06/19/science/studies-of-human-microbiome-yield-new-insights.html?_r=0

Experiment 12: Assess activity against eukaryotes, potential use in biological control, and ecological relationships with other organisms

Biological Questions:

1. Could we use the antibiotics produced by our isolates to treat bacterial infections in humans? Why or why not?
2. How do our antibiotic-producing isolates interact with plants and other organisms?
3. Could we use the antibiotic-producing organism as a biological control agent in our crops? Against what disease?

Isolate's interactions with plants:

Hypothesis and rationale:

Experimental design and protocols used:

Controls and treatments:

Observations:

Interpretations and conclusions:

Isolate's effect on plant pathogens, e.g., *Pythium*:

Hypothesis and rationale:

Experimental design and protocols used:

Controls and treatments:

Observations:

Interpretations and conclusions:

Design experiment to assess other ecological relationships:

Concluding Remarks

Think about how far you have come in only a few weeks. You have learned about the global antibiotic crisis, you have learned that soil is not just "dirt," you have isolated bacteria from your own soil samples, and you have applied the scientific method to investigate these samples. Throughout this time, you have learned how to communicate with your peers in your classroom and around the world and you have shared your experimental results with them. You are on your way to becoming an active partner in the educational mission of your campus and a lifelong learner.

Furthermore, you have played an important role in addressing what the United Nations has called "the greatest and most urgent global risk." While local, national, and global actions are required to solve the antibiotic crisis, individuals can have an impact. Going forward, consider simple things you can do to stop the misuse of antibiotics and reduce the spread of infections, including washing your hands, supporting companies that adopt safe antibiotic-use policies, getting vaccinated, and not demanding antibiotics when you have a viral infection like a cold or flu. Remember, in the United States, consumer demand led to nine of the largest food chains adopting new sourcing policies that require antibiotic-free meat, which have prompted changes in agricultural practices. It is important for you to understand your power as an individual and a consumer.

While we are running out of time to deal with the public health emergency of antibiotic resistance, it is not too late if we respond effectively with global collaboration. Through your research, you have already been part of a global initiative to crowdsource antibiotic discovery through the efforts of thousands of student researchers conducting mass-scale fundamental biological discovery.

Even though antimicrobial resistance has significant global consequences, humans have the ingenuity to solve this problem...if we act. We hope that everyone who is part of **Tiny Earth** recognizes that, together, we can make a difference.

PART II: Research Protocols

Contents

Agarose Gel Electrophoresis

Agarose gel electrophoresis is a common technique in molecular biology and microbiology that separates DNA fragments based on three features: (1) charge (positive versus negative), (2) size in base pairs (bp), and (3) conformation (linear versus circular). Scientists use agarose gel electrophoresis to follow up on DNA extraction and amplification protocols, such as polymerase chain reaction (PCR), which generate millions of DNA fragments. Running a DNA solution or reaction through an agarose gel separates millions of DNA fragments into discrete bands that indicate size, quantity, and uniformity.

An agarose gel is a jellylike material made up of agarose, a polysaccharide extracted from seaweed, and a buffer solution. Agarose forms a porous matrix that allows molecules to pass through with some impedance determined by the concentration of the agarose and size of the molecules. When the gel is placed in an electric field in buffer solution, charged molecules migrate through the gel toward the opposite charge. One such molecule is DNA, which has a partial negative charge on the phosphate group of the DNA backbone. DNA molecules loaded into a gel migrate toward the positive electrode. Since the gel matrix impedes flow, smaller molecules will travel rapidly, easily slipping through pores in the matrix, whereas larger molecules will be slowed down as they bump into the gel matrix; this process spreads the DNA molecules through the gel based on their size.

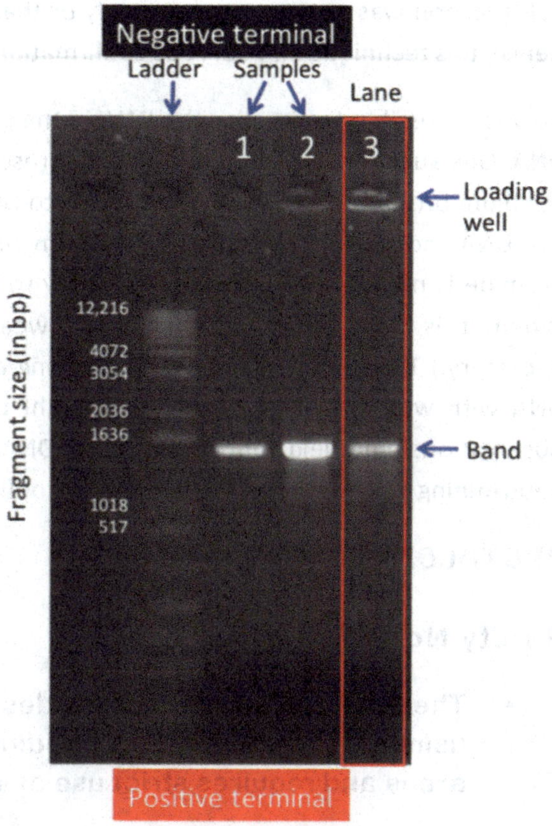

Figure 1. An agarose gel with a 1-kb ladder reference. A 1-kb ladder consists of fragments of DNA that differ in size by 1000-bp (1-kb) increments. It is used as a standard for comparison with unknown samples. In this gel, the 1-kb ladder is in the left-most lane and the other three lanes contain samples with discrete bands that represent DNA molecules that are approximately 1500 bp. DNA migrates through the gel in a buffer solution from the negative terminal to the positive terminal when an electric field is applied.

A PCR product, which contains millions of copies of one specific stretch of DNA, will appear as one discrete band on a gel. A solution containing a mixture of DNA fragments of different sizes will generate either multiple discrete bands or a smear of randomly sized fragments. To estimate the size of the fragment in each band, a molecular-weight size marker or DNA ladder, which contains DNA fragments of known sizes, is run in its own lane on the same gel with the unknown samples. This way, scientists can use gels to analyze mixed samples or PCR products, given that they have a basic understanding of the DNA sample or product at hand. Amplifying regions of bacterial DNA, such as the 16S rRNA gene, with PCR will generate fragments of approximately 1500 bp or smaller. Therefore, we would expect to see one band that has traveled the same distance as a 1500-bp band on the ladder. Any other result, such as a missing band, multiple bands, or a smear, may indicate that the original PCR reaction was not executed properly or that the template DNA did not contain the desired template. Hence, this technique may serve as confirmation that the predicted product was synthesized.

How is it possible to visualize the DNA in the gel? Various fluorescent tags emit light when they bind to DNA. One such tag, ethidium bromide, fluoresces with an orange color when exposed to ultraviolet light. Ethidium bromide molecules intercalate into DNA, or insert between adjacent base pairs, which disrupts the DNA molecule. In living cells ethidium bromide can produce mutations. (Note: Use of ethidium bromide is minimized because of its ability to produce mutations, which can lead to cancer in animals. When it is used precautions including wearing proper personal protective [PPE] equipment are necessary.) The brightness of the fluorescing ethidium bromide is indicative of the relative quantity of DNA with which it is associated — a bright band contains more DNA than a dim band. While these observations only yield approximations of DNA size and concentration, they serve as a useful first step in determining the outcome of a reaction or confirming the presence of a specific product.

PROTOCOL

Safety Notes:

- **The protocol used here provides guidelines for agarose gel electrophoresis using ethidium bromide.* Ethidium bromide should be handled only in designated areas and requires strict use of lab coats, gloves, and goggles.**

*Substitutes for ethidium bromide include SYBR safe and GelRed, which is more sensitive and stable than either ethidium bromide or SYBR safe.

Note: Different tags and stains may require different viewing trays and safety precautions. Chamber and gel tray sizes may also vary depending on the equipment available at your facility and the number of samples to be analyzed.

MATERIALS

- Agarose
- 1× TBE (TRIS-borate-EDTA) buffer
- 1-kb ladder
- Ethidium bromide (1% solution in deionized [DI] water)
- Erlenmeyer flask
- Microwave oven
- 6× loading dye
- Gel tray, comb, chamber, and power pack for electrophoresis
- UV chamber or lamp

PROTOCOL

1. Weigh 1 g of agarose and mix with 100 mL 1× TBE buffer in an Erlenmeyer flask – this will make a 1% agarose gel
2. Microwave contents of Erlenmeyer flask (1-2 min) until the mixture boils or becomes transparent.
3. Allow mixture to cool (10–15 min)
4. Carefully add 2–3 µL of ethidium bromide to cool agarose mixture. Pour mixture into gel tray with appropriate comb and allow 20 minutes to solidify.
5. Carefully remove comb from gel. Place gel tray into gel electrophoresis chamber, submerged in 1× TBE buffer, and start loading wells:
    a. Load 1-kb ladder into first well.
    b. Mix 5 µL of PCR product with 1 µL 6× loading dye and dispense into wells.
6. Plug electrodes into appropriate terminals of the chamber (positive with positive, negative with negative), ensuring that DNA is loaded on the negative end of the gel. What will happen if you plug in the electrodes backward?
7. Allow gel to run for 40–60 minutes (at 100 mA/V or constant voltage at approximately 180 V, depending on the equipment).
8. Finally, carefully remove gel from tray and place in UV chamber to photograph it (expect to see band at about 1.5 kb for full-length 16S rRNA gene).

Antibiotic Resistance Test

Note alternative method at the end of this protocol.

Soon after the introduction of antibiotics into clinical practice, resistant microorganisms were detected. Surprisingly, resistance to synthetic antibiotics (created by chemists rather than found in Nature) appeared as rapidly as resistance to antibiotics found in Nature, proving incorrect predictions that because the synthetic antibiotics didn't come from Nature, resistance hadn't yet evolved. Antibiotics that take tremendous resources to develop for clinical use can become ineffective within months or years of their introduction – often due to drug misuse or overuse. The tendency of antibiotics to stop working has deterred pharmaceutical companies from pursuing new antibiotics even though the demand continues to rise. So, how does antibiotic resistance come about? We have noticed that, generally, low- and long-term exposure to antibiotics provide a selective pressure for microorganisms with favorable genotypes – those carrying resistance genes or with spontaneous mutations – to increase in frequency in the population (more on antibiotic resistance in Section 10). Selection pressure can result in proliferation of resistant bacteria at the expense of sensitive ones of the same species, and the resistance genes can even be transferred to other species through horizontal gene transfer. This evolutionary phenomenon has existed for as long as antibiotics have existed in Nature; however, the incredibly fast rate at which antibiotic resistance spreads is associated with mass production and widespread use of antibiotics and is problematic.

Resistance varies in distribution and mechanism. Soil bacteria are resistant to their own antibiotics, often because of genes that confer resistance only to the antibiotic produced. Other resistance genes provide resistance to several antibiotics, and some bacteria contain many antibiotic resistance genes, making them resistant to many antibiotics. The soil antibiotic resistome is the collection of resistance genes from microorganisms in the soil. While humans have vastly increased the frequency of resistance genes in the bacterial world, resistant bacteria were present before antibiotic use by humans, and resistance genes can be found even in the most remote environments on the planet.

In this protocol, we will test bacterial isolates for the presence of resistance genes against common antibiotics. ***For more information on antibiotic resistance, please refer to Section 10.***

MATERIALS

- Antibiotic-containing media plates (see preparation for instructor below)
- Sterile inoculating loops or toothpicks
- Streak plates of isolate to be tested or original master plate

Preparation of Antibiotic-Containing Media (for instructor)

1. Prepare LBA or isolate's original medium. Make enough media for each student to have one plate with medium of choice per antibiotic.
2. Place aliquots of media into separate capped bottles that will each be mixed with an aliquot of antibiotic solution. Placing the media into the bottles prior to autoclaving will decrease the

chance of contamination. Alternatively, use aseptic technique to distribute the media after autoclaving.

3. After autoclaving media, allow to cool down (comfortably warm to the touch). Be sure to work quickly to keep the media from solidifying, or keep it warm in a water bath.
4. Follow antibiotic specifications in the table below for the following steps.
5. Determine the volume to antibiotic stock solution to be added to the batch of media.
6. Add x volume of antibiotic solution to batch of media, cap bottle, and mix gently by swirling to distribute antibiotic solution.
7. Pour plates and label appropriately. Keep plate containing light-sensitive antibiotics wrapped in foil or in the dark. After solidifying, store plates until lab session.

Antibiotic Specifications

Antibiotic	Solvent	Stock solution and storage temperature	Working solution (final concentration)	Light sensitive?***	Antibiotic mode of action?
Penicillin	DI* water	10 mg/mL (–20°C)	10 µg/mL media	No	
Gramicidin	Methanol	10 mg/mL (–20°C)	10 µg/mL media	No	
Trimethoprim	DMSO**	5 mg/mL (–20°C)	5 µg/mL media	No	
Rifampicin	DMSO	10 mg/mL (–20°C)	1 µg/mL media	Yes	
Tetracycline	70% ethanol	10 mg/mL (–20°C)	1 µg/mL media	Yes	

*Deionized

**Dimethyl sulfoxide

***Media with light-sensitive antibiotics must be wrapped in foil or stored in the dark. Plates must also be wrapped in foil or stored in the dark.

PROTOCOL

Safety Notes:

- **Must wear gloves and goggles, and practice aseptic technique throughout this protocol.**
- **Handle all solvents in a chemical fume hood.**
- **Ensure proper glove choice with use of DMSO since it can be absorbed directly into the skin.**
- **Label all secondary containers of chemicals and do not allow DMSO to enter drains.**
- **Dispose of chemical waste according to your institution's policies.**

Obtain streak plates of your isolates (recommended), or original master plate.

1. Obtain antibiotic-containing media plate(s) of interest. Media plates will contain standard media (LB, PDA, TSA, etc.) with the following antibiotic concentrations:

   a. Penicillin: 10 µg/mL
   b. Gramicidin: 10 µg/mL
   c. Trimethoprim: 5 µg/mL
   d. Rifampicin: 1 µg/mL
   e. Tetracycline: 1 µg/mL

2. Review Table 1 with antibiotic specifications. Ensure that antibiotic-containing media plate is labeled appropriately with antibiotic, concentration, and isolate(s) to be tested.
3. Using sterile inoculating loop or toothpick, pick isolate(s) from streak plate(s) or master plate and patch onto antibiotic plate. Make sure patch is labeled on the back of the plate with isolate name.
4. Incubate plate at isolate's designated growth conditions for 1–2 days, or until dense patch forms. If necessary, ensure plates are refrigerated at 4°C until next lab session to prevent overgrowth of patch (if next lab session is more than 2 days later).
5. After incubation period, observe plates for growth. Does the isolate appear to be susceptible or resistant to the antibiotic? Develop a hypothesis that explains the result based on your observations and what you know about your isolate. Record observations.

NOTE ON PROTOCOL ALTERNATIVE: Instructor may use the Kirby-Bauer Disk Diffusion Susceptibility test as an alternative to this protocol. This technique may be of interest to aspiring microbiologists. For more details on the Kirby-Bauer protocol, and to see the procedure, please the ASM Microbe Library's Laboratory Protocols at https://www.microbelibrary.org/.

Catalase Test

Adapted from: Karen Reiner – Microbe Library http://microbelibrary.org/index.php/library/laboratory-test/3226-catalase-test-protocol

Hydrogen peroxide (H_2O_2) is a strong oxidizing agent that rapidly kills susceptible cells and is employed by many organisms to protect themselves against infectious bacteria. H_2O_2 is also used as an antiseptic applied on cuts or wounds on the skin; you will typically find it next to rubbing alcohol and first-aid kits at the store. Many bacteria have evolved defense mechanisms that allow them to counteract H_2O_2 activity in order to survive. This is facilitated by catalase, an enzyme that breaks down H_2O_2 into water and oxygen gas.

$$2H_2O_2 + \text{catalase} \rightarrow 2H_2O + O_2$$

Bacteria containing catalase will effervesce, or form bubbles of oxygen, when they come in contact with H_2O_2. This serves as a simple test for the presence of catalase. For this reason, it is one of the bacterial enzymes described and used to classify bacteria (Gagnon et al., 1959; McLeod et al., 1923). A 3% H_2O_2 solution is typically used to test bacteria for catalase activity. Other applications, especially in diagnostics, include the identification of anaerobes using 15% H_2O_2 solution, as anaerobes typically lack the enzyme (Bartelt, 2000).

The catalase test is essential for differentiating catalase-positive *Micrococcus* and *Staphylococcus* from catalase-negative *Streptococcus*. While it is primarily useful in differentiating among genera, the test is also valuable in speciation of certain Gram positive bacteria such as *Aerococcus urinae* (positive) from *Aerococcus viridans* (negative) and Gram-negative organisms such as *Campylobacter fetus*, *Campylobacter jejuni*, and *Campylobacter coli* (all positive) from other *Campylobacter* species (MacFaddin, 2000; Mahon et al., 2011). Some microbiologists use the amount of activity in the catalase test to differentiate among certain members of the *Enterobacteriaceae* family (Taylor and Achanzar, 1972). The catalase test is also valuable in differentiating aerobic and obligate anaerobic bacteria, as anaerobes are generally known to lack the enzyme (Mahon et al., 2011; McLeod et al., 1923). In this context, the catalase test is valuable in differentiating aerotolerant strains of *Clostridium*, which are catalase negative, from other members of the same phylum, such as *Bacillus*, which are catalase positive (Mahon et al., 2011).

MATERIALS

- 3% H_2O_2
- glass microscope slides or petri dishes
- inoculating loop or toothpicks
- streak plates of isolates
- streak plates of positive and negative controls (from safe ESKAPE relatives: Positive: *P. putida*; Negative: *S. epidermidis*

PROTOCOL

1. Obtain the streak plate of your bacterial isolate with fresh (1–3 day old), fully grown colonies.
2. Obtain a microscope slide or piece of parafilm. Optional: place microscope slide/parafilm piece inside a Petri dish and cover with lid to limit catalase aerosols during the test.
3. Using a sterile inoculating loop, stick, or toothpick, pick a single colony from the streak plate. A small yet visible cell mass will suffice. Note: Be careful not to pick any agar, especially if the isolate was grown on agar containing red blood cells, as these may result in a false-positive reaction.
4. Gently apply the cells with a back-and-forth motion until there is a small, visible smear in the center of the slide/film. To conserve materials, smear up to three isolates in separate smears on a single slide.
5. Using dropper or pipette, place 1 drop of 3% H_2O_2 solution onto the smear on the slide. Do not mix. Note: carrying out steps 5 and 6 on a dark background will enhance the visibility of bubbles.
6. Observe for the production of oxygen bubbles – this will happen rapidly and vigorously in a positive reaction. A weak reaction or no reaction may indicate a negative. Optional: For weak catalase-positive microorganisms, examine slides under a dissecting microscope or using 4x or 10x magnification on a light microscope.
7. Perform a control reaction using organisms known to be positive and negative for catalase.

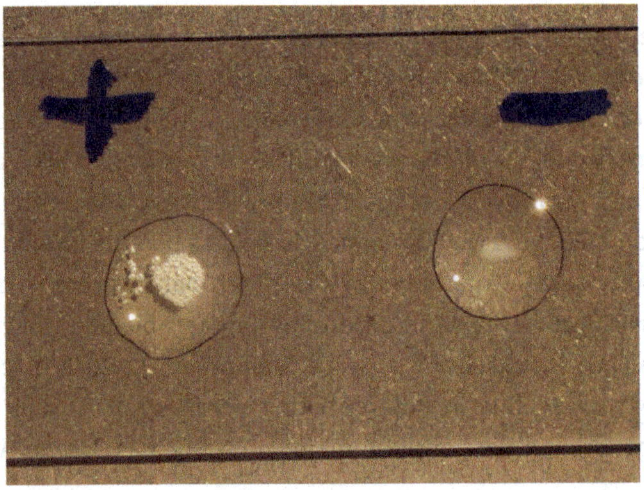

Figure 2. Catalase test. *Bacillus subtilis* is catalase positive (left) and *Enterococcus raffinosus* is catalase negative (right). Photo source: Manuel Fernando Garavito Diago

REFERENCES

Gagnon, M., Hunting, W., & Esselen, W. B. (1959) A new method for catalase determination. Anal Chem 31:144.

MacFaddin, J. F. (2000) *Biochemical Tests for Identification of Medical Bacteria* (3rd ed.). Philadelphia, PA: Lippincott Williams & Wilkins.

Mahon, C. R., Lehman, D. C., & Manuselis, G. (2011) *Textbook of Diagnostic Microbiology* (4th ed.). Philadelphia, PA: W. B Saunders Co.

McLeod, J. W., & Gordon, J. (1923) Catalase production and sensitiveness to hydrogen peroxide amongst bacteria: With a scheme for classification based on these properties. J Pathol Bacteriol 26:326–331.

Taylor, W. I., & Achanzar, D. (1972) Catalase test as an aid to the identification of Enterobacteriaceae. J Appl Microbiol 24: 58–61.

Adapted from:
Reiner, K. (2010) "Catalase Test Protocol". *Laboratory Protocols*. **Microbe Library**. American Society for Microbiology. Accessed 13 October 2013. http://microbelibrary.org/index.php/library/laboratory-test/3226-catalase-test-protocol

Chemical Extraction of Antibiotics

Separations have been a central theme in our research course: we have separated microbes from their natural habitat in our sample collections, we have separated individual cells in spread-and-streak plates, and now, our goal is to separate their bioactive compounds to be studied in isolation. This reductionist approach enables us to study cells, reactions, and metabolites under conditions that reduce confounding variables (other cells or molecules). Separating a bioactive compound from other molecules present in a culture is the first step in determining that compound's physical and chemical properties, mode of action, and activity spectrum. Just the way we use many household products to remove stains on fabrics and surfaces, we can apply specific solvents to cultures to extract compounds with similar properties. These compounds are typically classified based on their **polarity**, a chemical property discussed in Section 9. Polar or water-soluble compounds are at one end of the polarity spectrum while nonpolar or fat-soluble compounds lie at the opposite end. Chemists typically like to use solvents that have intermediate polarity to extract a compound of interest without getting all the "junk" at the extremes, such as water-soluble nutrients and long-chain fatty acids. In this protocol, we will use one of two common solvents, ethyl acetate or methanol, to attempt to separate the bioactive compound from most cell components and nutrients in our cultures. Whether the bioactivity is retained in water and/or ethyl acetate indicates the polarity of the active molecule and suggests ways that the compound or compounds can be manipulated in future experiments. ***For more information on chemical extractions, please refer to Section 9.

SESSION 1 – PREPARING PLATES

MATERIALS

- One medium plate per isolate (same as isolate's original medium)
- Sterile plastic inoculating loop or sterile swab– 1 per isolate
- Streak plates of isolates (or original master plate)

PROTOCOL

Safety Notes:

- **Must wear gloves, lab coat, and goggles. Use sterile technique in every step of Session 1 Protocol.**

1. Select isolates to be cultured for chemical extraction (1 or 2 per student)
2. Obtain streak plate of isolates, or original master plate if streak plates are missing.
3. Using the large end of the sterile loop, collect enough single colonies of isolate to spread on the entire surface of a media plate (fill about half of the loop). Or use a sterile swap to pick up a few colonies of isolate to spread across the surface of a media plate.
4. Gently inoculate entire surface of the fresh media plate by rubbing the loop or swab with colonies back and forth until the entire area is covered.

5. Ensure the plate is evenly coated with inoculum. If needed, turn the plate 90° and continue to rub the loop/swab back and forth. This will grow into a **lawn** of your isolate.
6. Incubate plates at 28°C, or temperatures and conditions specific to isolate. Incubate until next lab session, or 2-3 days, and up to a week.

SESSION 2 – CHOPPING UP PLATE

MATERIALS

- 100-250-mL glass bottles (with cap) – 1 per isolate
- Microspatula or cutting device
- Freezing compartment*

*Options include a freezer (–20°C or –80°C) or a dry ice/ethanol bath

PROTOCOL

Safety Notes:

- **Must wear gloves, lab coat, and goggles. No need to keep lawn sterile with a Bunsen burner or flame in Session 2 Protocol.**

1. Obtain inoculated media plates from previous session. Ensure your isolates have grown into a dense lawn.
2. Using a microspatula, cut the media plate with fully grown isolate into small pieces, about 1 cm^2 each (no need to make precise cuts).
3. Scoop all the pieces with microspatula into a 100-250-mL glass bottle and label with isolate name. Push all the pieces to the bottom of bottle with microspatula.
4. Freeze bottle in freezing compartments. Follow specific instructions for freezer compartment of choice below:

   FOR FREEZER, place bottle in freezer until next lab session (CONTINUE ON TO **SESSION 3-A** IN THE SUBSEQUENT LAB PERIOD)

   FOR DRY ICE/ETHANOL BATH, place bottle into bath for 20–30 minutes (Caution: ethanol with dry ice is supercooled to temperatures that can cause frostbite upon contact with skin. Do not splash ethanol. Place bottles into bath gently to keep glass from shattering!) (CONTINUE TO **SESSION 3-A** IN THE SAME LAB PERIOD)

Question: What is the purpose of freezing the pieces of agar?

SESSION 3-A – ORGANIC EXTRACTION PART 1

MATERIALS

- Ethyl acetate **or methanol**
- Deionized water **(not recommended for methanol extraction)**

- Pasteur pipettes (glass) and bulbs
- 20-mL scintillation vial or other glass container* – 2 per isolate (1 per isolate for methonal)
- Analytical scale (optional)
- Fume hood
- Recommended: drying-system or vacuum

PROTOCOL

Safety Notes:

- **As a general precaution, always use glassware instead of plastic when working with solvents.**
- **Must wear gloves, lab coat, and goggles. Must handle solvents inside fume hood.**
- **Inhalation of organic solvents can cause adverse health effects, drowsiness, and loss of consciousness.**
- **Keep away from Bunsen burner or flame when working with open bottles or beakers with flammable solvents, such as ethyl acetate and methanol.**

Note: All steps on this protocol apply to both ethyl acetate and methanol, unless otherwise noted. Ethyl acetate-specific instructions are underlined. **Methanol-specific instructions are bold and in red**. Pay close attention to differences in the text.

Note: No need to keep things sterile with Bunsen burner or flame in Session 3 Protocol.

1. Remove bottles from freezer OR dry ice/ethanol bath. Bottle may be thawed or kept frozen for next step.
2. FOR ETHYL ACETATE EXTRACTION: Add 15 mL of ethyl acetate and 10 mL of water to bottle. You may notice two layers will start to form. The top layer is the ethyl acetate phase, and the bottom layer is the aqueous phase.
   What does the placement of layer say about the density of ethyl acetate?
   Place capped bottle on shaker at room temperature until next lab session.

   OR

   FOR METHANOL EXTRACTION: Add 15 mL of methanol (DO NOT ADD WATER) to bottle and place capped bottle on shaker at room temperature until next lab session.

   Note: Methanol and water are miscible, which means they can mix together. Mixing methanol with water will dilute our extract and will make it difficult and time-consuming to dry the extract.

While shaking at room temperature, bottle will thaw and organic solvent will extract soluble components in the lawn and agar.

SESSION 3-B – ORGANIC EXTRACTION PART 2

1. Using a Pasteur pipette and bulb, gently transfer all the liquid from the bottle to a labeled scintillation vial or other glass container. Once all liquid is transferred, let sit for 2–3 minutes to allow any debris to settle (e.g., small chunks of agar).

   FOR ETHYL ACETATE, organic (top) and aqueous (bottom) layers will start to separate. Depending on the color of your extract, both layers may appear clear. Carefully identify the interface between them. Note: Emulsion (a cloudy third layer) may form between the top and bottom layers. Try not to disrupt the emulsion.

   FOR METHANOL, note that separate layers will not form. If pre-weighing vial, only one vial is needed and should be weighed prior to transfer step.

2. Optional*: Pre-weigh second set of scintillation vials or other glass container with appropriate labels (isolate name, etc.) on analytical scale. Record weights.
3. Using a Pasteur pipette and bulb, transfer solvent from vials or containers in step 3 to preweighed vial or container. Follow specific instructions for solvent of choice below:

   FOR ETHYL ACETATE, transfer organic layer (top) – ethyl acetate with soluble components of the lawn and the agar – into pre-weighed vial or container. Carefully collect as much of the organic layer as possible. Avoid collecting emulsion (if present) and aqueous layer. It is better to leave some solvent than collect water. Once organic layer is removed, keep vial or container uncapped in fume hood to evaporate remaining ethyl acetate.

   FOR METHANOL, all 15 ml transferred to vial should be placed uncapped in fume hood to evaporate solvent.

   After that, store aqueous phase at 4°C or freeze at –20°C if intention to assay aqueous phase for activity (refer to note at the end of Session 4 Protocol).

4. Allow extract to dry down by evaporation. Vials or containers with extracts may be left uncapped inside fume hood until next lab session, or until completely dried down.* Alternatively, extracts may be dried using drying system or applied vacuum, which may reduce the drying-down period to several minutes or hours.

*Pre-weighing empty vials and containers prior to transferring extracts, and then re-weighing with dried extract, makes it possible to calculate how much extract is recovered from cultures. This facilitates preparing extracts at the same concentration, which makes it possible to compare the potency of different samples by assaying their activity based on a known concentration. Knowing the concentration of material facilitates using appropriate amounts of material in subsequent chemical analyses *via* Liquid Chromatography Mass Spectrometry or Thin Layer Chromatography (TLC). An analytical balance is highly recommended because the weight of extracted material can range from milligrams to submilligrams (less than 10^{-3} grams). It is advisable to keep the same vial and lid together since lid weights can vary substantially and a lid from different extract can contaminate a sample.

SESSION 4 – ASSAY EXTRACT FOR ANTIBIOTIC ACTIVITY

Note: All dried extracts will be redissolved in methanol, regardless of the solvent used in the extraction – Session 3.

MATERIALS

- Dried-down extract in scintillation vial or other glass container
- Media plates (because we are testing cell-free extract, does not need to be the same as isolate's original medium)
- Overnight culture of safe ESKAPE relative (your tester strain)
- Methanol
- Sterile 15-mL conical tube or test tube
- Micropipettor and tips (P20 and P200)
- LB with 1/2x (0.75%) agar (if using top agar method)

PROTOCOL

Safety Notes:

- **Must wear gloves, lab coat, and goggles, and use sterile technique.**

1. Your extracts are now dried down in a vial/container.
   Optional (if vials/containers were pre-weighed): re-weigh vial/container on the same analytical balance, calculate mass of dried extract, and record in your notes.
2. IF MASS NOT CALCULATED: Resuspend dried-extract in 80 µL methanol

   OR

   IF MASS CALCULATED: Resuspend pre-weighed dried-extract methanol – adjust volume of solvent to attain desired concentration (recommended concentration: 5–10 µg/µL).

Spread-Spot Method

1. Spread some of an overnight liquid culture of safe ESKAPE relative (tester strain) using sterile loop or sterile swab on media that best supports tester strain growth (generally LBA or 1/10 TSA). Allow spread tester strain to soak into media so plate is dry for next step. Initially you should use the same tester strain(s) your isolate had activity against.

2. Prepare to spot dried extract on a spread plate of safe ESKAPE relative (tester strain):

   Indicate on your media plate where extract and control will be spotted by drawing a dot on the back of the plate (one circle for each extract and one for the control).

   Note: Properly label back of media plate with safe ESKAPE relative name, isolate name, and extracts and controls to be spotted into the circles.

Question: What will the control(s) be in the extract activity assay?

1. Using a P20 micropipettor, spot 10 µL of resuspended extract on the appropriate pre-drawn dot. Allow the solvent to evaporate – the spot should appear dry. Repeat for methanol control.
2. Load another 10 µL on same spot and let dry (OR volume determined for specific concentration). Repeat for the control.
3. Repeat step 5 one more time until 30 µL total volume is loaded onto the appropriate pre-drawn dot.

Top Agar Method

***Note: Top agar (LB with 1/2x (0.75%) agar) must be liquefied in microwave or water bath and allowed to cool down (warm to the touch) prior to Step 8 below. If cooled for too much, it will solidify.

3b. Skip step 3 above, but continue with steps 4-7, spotting extract and control on sterile media plate. Because we are using top agar (which supports growth of tester strain) and cell-free extracts the type of media is not as critical here.

While the spots are drying, obtain an overnight liquid culture of your tester strain and an empty, sterile conical tube or test tube. Set P200 micropipette to 100 µL.

***Following steps must be done carefully but quickly to keep liquefied top agar from solidifying in the tube. Following steps will not be used for Spread-Spot method. ***

1. Using sterile technique, transfer ~7 mL of liquefied top agar to the sterile tube.
2. Inoculate liquefied top agar in tube with 100 µL of tester strain overnight liquid culture. Gently mix by rolling tube back and forth in hands.
3. Quickly and carefully pour inoculated top agar onto spotted media plate. Gently tilt the plate to spread top agar evenly over the entire plate. After covering entire surface, keep plate upright until top agar solidifies (i.e., becoming translucent and appearing more rigid).
4. Incubate at 28°C, or temperatures and conditions specific to tester strain. Plate may be incubated upside-down after top agar solidifies. Incubate overnight or until tester strain grows into a lawn, and keep refrigerated at 4°C to avoid overgrowth until next lab session.

AQUEOUS PHASE NOTE: Session 4 Protocol may be repeated using aqueous phase. Note that aqueous spots will take significantly longer than those in organic solvents to evaporate in the fume hood and may not fully dry in one lab session. Alternatively, we recommend using a lyophilizer to dry down the frozen aqueous phase and then proceeding through Session 4 Protocol.

SESSION 5 – SCREEN ZONES OF INHIBITION

MATERIALS

- Plates from previous session

PROTOCOL

Safety Notes:

- **Must wear gloves, lab coat, and goggles, and use sterile technique.**

Observe plates for zones of inhibitions. Ensure that methanol control worked properly. Measure zones of inhibition and record observations. Did your organic extracts retain antibiotic activity? How do the zones of inhibition compare with observations made in previous assays?

Alternatives: EXTRACTION WITH OTHER SOLVENTS

2-Butanol:

Although 2-butanol is immiscible with water, the organic and aqueous fractions are typically not as distinct as when ethyl acetate is used. The cloudy region in between these layers can be eliminated using centrifugation. Transfer the liquid to a conical tube and centrifuge for 5 minutes at 10,000 rpm. The organic layer can then be easily pipetted to a scintillation vial for drying.

Hexane:

Hexane is easy to use for extractions because it evaporates very quickly. However, when hexane is spotted onto agar, it spreads across the plate quite far from the point where it was originally spotted. Therefore, if hexane is used as the extraction solvent, the bioassay procedure must be altered. Rather than spotting 10 µL of re-suspended extract 3 times (as described in "Day 4"), spot 5 µL of re-suspended hexane extract 6 times.

Colony Morphology Description

Since bacteria were first cultured on solid media, microbiologists have noticed that colonies display differences in physical appearance at the macroscopic level. Reference books such as *Bergey's Manual of Systematic Bacteriology* contain extensive descriptions of bacterial physical and functional attributes, which have enabled microbiologists to identify and classify bacteria for nearly a century. Colony morphology is one of the first things we observe about bacteria at the macroscale and use to distinguish them from other microbes and distantly related bacteria. The main descriptors used by microbiologists are: (1) color, (2) surface texture and elevation, (3) shape, and (4) margin. We use specific terms to describe morphological characteristics, such the ones described in Figure 3. However, bacteria come in a many forms; therefore, there is much room for creativity when describing colony morphology.

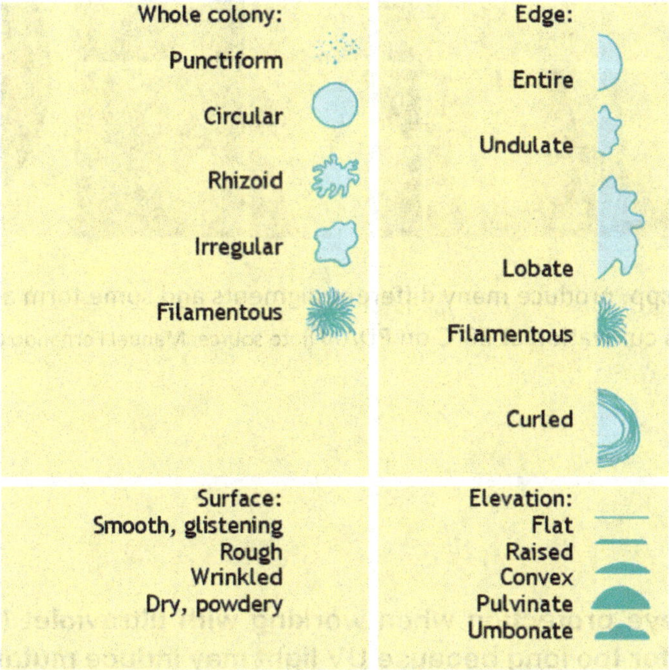

Figure 3. Descriptors of colony morphology. Photo from commons.wikimedia.org

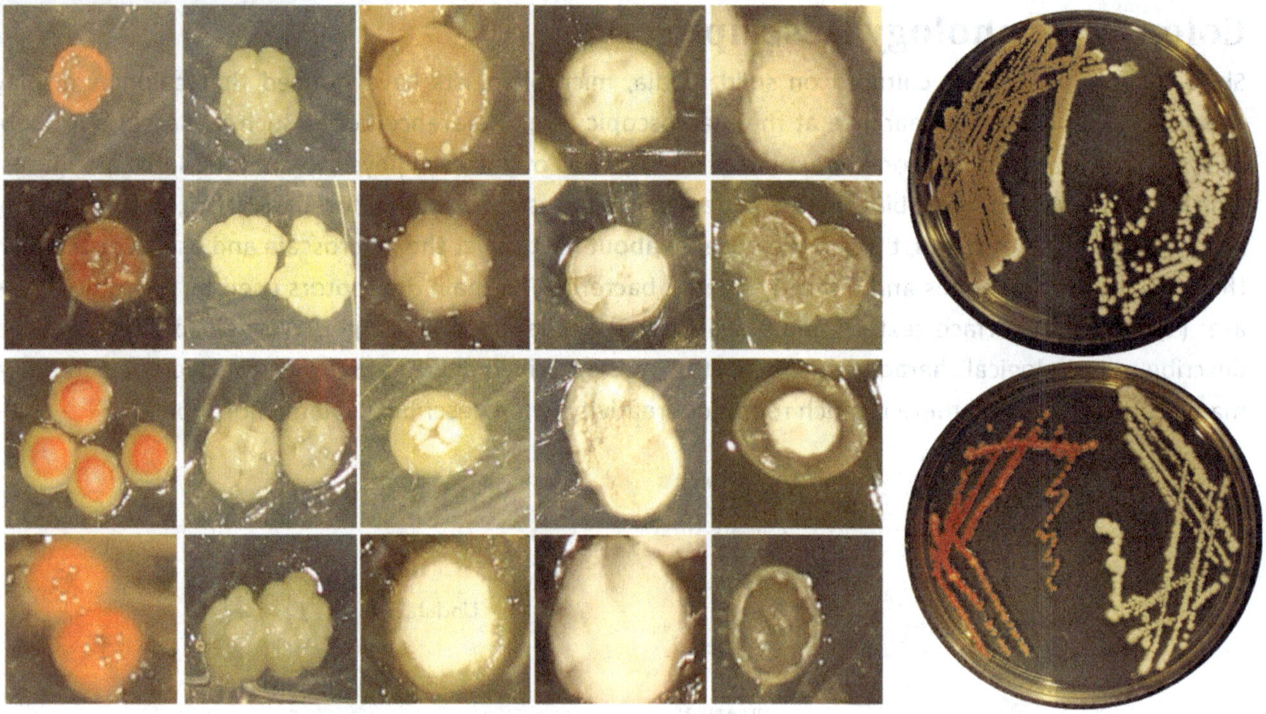

Figure 4. *Streptomyces* spp. produce many different pigments and some form aerial mycelia. Soil isolates are shown after 10 days cultivation at 28°C on PDA. Photo source: Manuel Fernando Garavito Diago

PROTOCOL

Safety Notes:

- **Use proper eye protection when working with ultraviolet (UV) light, and do not expose cells for too long because UV light may induce mutations.**

1. Many classrooms will use colony morphology differences as the first level of selection from dilution plates to identify different isolates for further testing. Observe colonies from soil dilution series. Alternatively, allow streak plate of bacterial isolates or master plate to grow for 2-4 days under optimal growth conditions. **Note:** Even if colonies are visible after a one-day incubation period, give the cells more time to produce pigments and other morphological identifiers that may not be as evident during logarithmic growth phase.
2. Once the cells are in the stationary phase of growth, start observing and describing their physical characteristics. **Note:** Keep the plate covered whenever possible, and keep it covered and sealed with parafilm at all times when working with unknown isolates in a BSL-1 lab. **Optional:** Use a dissecting microscope for better observations.

3. Start by describing colony color, or pigmentation. Differentiate between colony pigments and pigments secreted into the agar surrounding the colony. If an ultraviolet lamp is available, shine the light on the colonies for few seconds to see if they glow in the dark.
4. Record an estimate of colony size; if possible, use a clear ruler to measure the diameter.
5. Pay close attention to the way colonies are shaped and scattered. Make sure you base your observations on single colonies. If colonies are too close together or packed together as a lawn or in slime, be sure to specify that in your notes. Note: You are not limited to using the terms in Figure 3.
6. Describe the margin, or edges, and elevation of the colonies. Although colony shape and margin influence one another, do not assume they are the same thing. **Note**: Carefully tilt the plate when observing colony elevation but do not bring it too close to your face.
7. Describe colony texture; this is where you will probably have to be most creative.
8. If possible, take pictures of your isolates using a lab camera. If your instructor allows personal devices in lab for picture taking, make sure it is protected in a plastic ziploc bag or wrapped in parafilm. Refer to Best Practice guidelines at the beginning of this book for more details.

Combining 16S rRNA Gene Sequences into a Single Contig

To determine an isolate's genus and closest matches in public databases, students can use PCR sequencing of the 16S rRNA gene and NIH's Basic Local Alignment Search Tool (BLAST) or the Ribosomal Database Project. Some classes may choose to sequence the full length (~1500 bases) of the 16S rRNA gene, which requires aligning the sequences from several PCR and sequencing reactions.

To obtain a preliminary identification of isolates, a single forward sequence can be used following the "DNA Sequence Analysis with BLAST and RDP" protocol. First, follow one of the "PCR Analysis" protocols, lysing cells and amplifying the 16S rRNA gene, verify product by gel electrophoresis and either purify the PCR product or pay for the sequencing facility to do the clean-up for you. We recommend sequencing using the forward (27F) primer. The 5′ end of the gene tends to be more variable and is more reliable in identifying the genus than using the reverse primer (1492R or 1391R). Alternatively, one can use a reverse primer from the middle of the gene (519R or 769R). Your preliminary analysis will usually provide you with between 500-750 bp of coverage of the 16S rRNA. However, you may want to obtain full-length sequencing coverage of the gene to better confirm the identity of an isolate. To do so additional regions of the 16S rRNA gene, both in the forward and reverse direction, should be performed. To do this, use the remainder of your initial product from the "Colony PCR" or repeat the procedure for 16S rRNA PCR amplification. At the sequencing step, prepare a sample for each primer to be used covering different regions of the gene. Some good options to start with are 27F, 515F, 769R, 1492R. Please visit www.tinyearth.wisc.edu to access the "PCR and Sequencing Troubleshooting Guide" for help in addressing common reasons for unsuccessful PCR, sequences for additional primers mentioned above, and suggestions for accurate interpretation of sequencing results.

After sending these samples to a sequencing facility and obtaining the results, trim the sequences as described above. You can then use several online tools to align these sequences into a single contig.

First, in order for the forward and reverse sequences to align, enter each of the three reverse sequences into the revseq tool (http://emboss.bioinformatics.nl/cgi-bin/emboss/revseq), selecting "Yes" for both the "Reverse sequence?" and "Complement sequence?" options. The results should be an inversion of the original sequence and each base from the original sequence should be replaced by its complement (G becomes C, C becomes G, etc.).

You can align sequences to create a consensus sequence. For this process, several programs can be used (see https://www.ebi.ac.uk/Tools/msa/). In addition, BLAST permits alignment of two sequences by selecting "Align two or more sequences" in the top box below where one inputs the query sequence. The results page will show an alignment of the sequences, which can be saved as a consensus file. Repeat this process for the three reverse sequences, which should provide two consensus sequences, one for the forward direction and one for the reverse direction. These can be combined into one final consensus sequence using the same procedure described above. You will likely find that the forward and reverse sequences have a large region of nearly perfect consensus in the middle and that the forward has additional bases at the beginning of the sequence, while the reverse has additional bases at the end of the sequence. To combine these two sequences, simply copy the extra bases present at the beginning of the forward sequence and paste them at the beginning of the reverse sequence.

Submit this final consensus sequence to BLAST using the above procedure. If the sequence is at least 97% identical to a species in the database, then you have identified the genus and species of your isolate. If your consensus sequence is less than 97% identical to the first BLAST result, then it is possible that your isolate is of a previously undiscovered species.

Culturing from Soil Sample

We will use a culture-dependent approach to study bacteria, which relies on the ability of some bacteria to grow in the lab. Microbiologists have used this approach for over a century, and while the premise has not changed, our increased understanding of bacterial nutrition and growth has enabled development of media and lab conditions to more closely meet bacterial requirements for survival. This has also increased the reproducibility of natural phenomena from lab to lab, enabling microbiologists to repeat each others' experiments and confirm observations, thereby advancing the field. While we are far from attaining perfection in culturing bacteria and accurately replicating their natural habitats, we have the tools of observation, inquiry, and ingenuity to come a step closer. In addition, the more we learn about environments like the soil, the more we realize that this is truly an inexhaustible source of bacteria from which we still have much to learn.

After you collect a soil sample, the first step is to suspend the bacteria in liquid so that they can be transferred to another medium (e.g., a nutrient plate), leaving behind plant material, soil animals, and minerals from the soil. It is important to take into consideration the kind of vegetation the soil supported and select appropriate media accordingly. The soil contains a wide range of bacteria, so it is likely to yield growth in almost any standard bacteriological medium.

MATERIALS

- 3 LB plates/group
- Conical tube
- Soil sample
- Spread beads
- Sterile water

PROTOCOL

1. Weigh 1 g of soil sample, and place into conical tube.
2. Determine 3 different ways to plate this sample to visualize microbes.
3. Record all three procedures in lab notebook.

DNA Sequence Analysis with BLAST and RDP

Genomes contain large amounts of information – the human genome is over 3 billion base pairs long and bacterial genomes like that of *Escherichia coli* contain nearly 5 million base pairs. Making sense of all this information requires biological knowledge of how genes work and what they encode and powerful computational tools to sort through large amounts of information and find patterns. Sequenced genomes are submitted into large databases like GenBank that are impossible to navigate without search tools, just as Google Search helps find specific information in the whole of the World Wide Web.

The Basic Local Alignment Search Tool (BLAST) (Altschul et al., 1990) is a bio-informatics tool that allows us to navigate through huge databases and compare an amino acid or nucleotide sequence with a library of sequences. BLAST comparisons of DNA sequences can find closely related genes or regions of DNA in the database. These closely related genes whose functions or organism of origin have been determined previously provide information about the likely function of a newly discovered gene's product or the identity of the organism from which it originated.

MATERIALS

- 16S rRNA gene* sequence and chromatogram
- BLAST website

*For more information on the 16S rRNA gene, please refer to Section 8.

PROTOCOL

1. Go to the BLAST website: http://www.ncbi.nlm.nih.gov/BLAST/
2. Choose BLAST program to run a "nucleotide blast"
3. Enter your sequence* into the "Enter query sequence" field and enter appropriate nucleotide range*
4. Under "Choose Search Set," select the "16S ribosomal RNA sequences (Bacteria and Archaea)" database on the drop-down menu to conduct a search.
5. Once you have submitted your sequence and set the parameters, click "BLAST" at the bottom of the page. BLAST will take a couple of seconds to compare your sequence to other sequences in the database.
6. Once the search is done, analyze your BLAST data. The "Descriptions" column lists identifiers for similar sequences in the database. These identifiers are ranked by "max ident," which is the percentage of matching nucleotides. Normally, a "max ident" of 97% or higher means that your sequence matches the specific description, which corresponds to a strain or species. The expectation value (E value) tells you how statistically significant your match is; hence, the lower the E value, the more reliable the match.**

*Before entering your sequence into the "Enter Query Field," be sure to assess the quality of your sequence. This can be done by looking at your sequence trace chromatogram, which shows the fluorescence peaks given off by each of the four nucleotides during Sanger sequencing. You can view and trim sequence chromatograms using freely available software (e.g., Finch, DNA Baser, Chromas, and 4Peaks). Peaks are considered low quality when they are not discrete or separated from one another; this usually happens at the beginning and end of the sequence. You can trim low-quality data from the beginning and end of the sequence or pick a "clean" nucleotide range in your sequence on which to conduct the BLAST analysis. After pasting your full sequence into the appropriate box, enter the appropriate region range into the "Query Subrange" fields.

**There are two numbers you should care about: Max score and E value. The higher the Max score, the better the match. The default BLAST algorithm assigns +2 for each matching nucleotide and –3 for each mismatch to give a raw score, and then adjusts that score by length and database size to calculate the "Bit score." The E value (or expected value) estimates how many times we expect to find a match of the same quality (bit score) in the database purely by chance. The lower the E value, the better. The matches with an E value above 1 can be discarded.

The Alignment - The first line with ">" gives the address (accession number) of the full entry in the database, name of entry, and length. The summary section below the name shows the bit score and raw score in parentheses, E value, and % identity. The last part is the alignment itself: the query on top and the subject (sequence from the database) on the bottom, and vertical lines for each match between them. The numbers specify nucleotide position in each sequence.

NOTE ON PROTOCOL ALTERNATIVE: Students may also input their sequences into the RDP database (http://rdp.cme.msu.edu/seqmatch/seqmatch_intro.jsp), which specifically contains curated bacterial and archaeal 16S rRNA gene sequences. Comparing the results from BLAST and RDP can be a simple way for students to confirm the identity of their sequence.

REFERENCES

Altschul, S. F., Gish, W., Miller, W., Myers, E. W., & Lipman, D. J. (1990) Basic local alignment search tool. J Mol Biol 215:403–410. doi: 10.1016/S0022-2836(05)80360-2

Fermentation Test

Adapted from: Clifford Grimsley – Gaston College

Many organic compounds can be produced through fermentation, and many microbes satisfy their energy requirements through fermentation in which carbohydrates are incompletely oxidized because oxygen is insufficient. This incomplete oxidation produces several different end products, such as acids, alcohols, aldehydes, and gases, such as methane or CO_2. Microbes ferment different substances and produce diverse end products, which are also characteristic of taxonomic groups; thus both the substrates and end products of fermentation are characteristic of taxonomic groups, making fermentation capabilities extraordinarily useful in microbe identification.

Fermentation of carbohydrates often results in the production of acid and gas. To determine whether an acid is produced, a pH indicator can be added to the growth medium. Growth media containing the pH indicator Phenol Red are normally red or pink at a pH 7, and a change to yellow indicates acid is produced. The production of gas is observed by using a Durham tube, which is an ordinary test tube that contains a small inverted vial. If gas is produced, the inverted vial will collect some of the gas and a bubble will be visible. The fermentation media are prepared by adding 0.05% of a single carbohydrate and 0.02% of a pH indicator to a nutrient broth medium. If the organism can break down the particular carbohydrate in the tube, then the various end products will be produced.

MATERIALS

- Durham fermentation tube containing lactose + phenol red
- Durham fermentation tube containing dextrose + phenol red
- Durham fermentation tube containing sucrose + phenol red
- Broth culture of controls.
- Streak plate of isolate

PROTOCOL

1. Pick bacterial isolate with a stick or inoculating loop. Inoculate the three carbohydrate broths with your isolate. Repeat for each isolate and control.
2. Be sure that each tube is labeled with the carbohydrate and the inoculated organism. Place the tubes in a beaker labeled with the initials of your group and the date.
3. Incubate all tubes at 37°C for 24–48 hours.
4. Your instructor will set up controls for you to view.
5. After incubation, compare each of the inoculated tubes with the control tube of the same media to determine whether acid or gas was produced. Record your results.

Fermentation Protocol Data Sheet

Sugar	***Isolate 1***	***Isolate 2***	***Isolate 3***	***positive control***	***negative control***
Dextrose					
Lactose					
Sucrose					

Record results as follows:

AG = Acid and gas produced
A= Acid only produced
a = Slight acidity produced
O = Carbohydrate not fermented

Glycerol Stock Preparation

When working with bacterial isolates in multiple experiments, it is important to preserve and store bacteria in an unchanging state that can be accessed for future use. Freezing allows us to keep bacteria in a dormant state for long periods of time, with one disadvantage – water forms ice crystals upon freezing, which can rupture and kill cells. To keep this from happening, we freeze bacteria in a glycerol solution, which bathes and stabilizes bacterial cells by preventing the formation of ice crystals. Bacteria or bacterial spores can be stored for years at –80°C in 20% glycerol solution and liquid medium.

MATERIALS

- 80% glycerol solution
- Fresh streak plate
- Liquid medium (same as streak plate)
- Cryogenic vials (1.5 or 1.8 mL)

PROTOCOL

Safety Notes:

- **Must wear gloves and goggles, and practice aseptic technique throughout this protocol.**

1. Streak out bacterial isoate on appropriate medium. Incubate until colonies are visible.
2. Pre-mix 80% glycerol and liquid medium in a conical tube. Adjust to a final concentration of 20% glycerol with liquid medium. (Note: if your isolate is streaked on LB agar, mix glycerol with LB broth.)
3. Dispense 1000–1200 µL of 20% glycerol mix into cryogenic vial. Label tube with isolate designation, glycerol concentration, and date. For example: SH-LB-1, 20% gly, 11/12/13. Use an alcohol-proof and waterproof marker.
4. Using a sterile inoculating loop, pick a single colony from the streak plate. Collect as much of the colony as possible without picking up agar.
5. Carefully pick the colony off the plate and quickly transfer it to the respective labeled cryogenic vial. Insert loop into the glycerol mix and gently wiggle or twist the handle to deposit the colony in the glycerol mix. ***Note: Whenever possible, keep the vial supported in a rack, and, if picked up for observation, keep below eye level and pointed away from your body.
6. Gently pipette up and down or gently vortex the cryogenic vial until the colony is uniformly mixed.
7. Freeze and store glycerol stock at –80°C.
8. Input the bacterial isolate information and location of glycerol stock in your notes and in the database.

Gram Stain Test

Gram stain employs two stains, crystal violet and safranin, which are violet and red, respectively. In Gram-positive bacteria, the peptidoglycan layer is very thick (>20 nanometers) and exposed to the outside. Gram-positive cells retain large amounts of crystal violet in this layer, and binding is fixed with a mordant, iodine, giving the cells a purple appearance under the microscope, even after washing cells with a decolorizer such as alcohol. Conversely, in Gram-negative bacteria, the peptidoglycan layer is thin (normally <10 nanometers) and covered by an outer membrane, the S-layer. It retains very little crystal violet but stains very well with the counterstain safranin, which gives the cells a pink appearance under the microscope. By discriminating between purple and pink cells on a microscope slide, clinical microbiologists can deduce much information about an isolate's physiology. Furthermore, Gram staining serves as a classification system for bacteria. Gram-positives and Gram-negatives make up two major subdivisions of bacteria, which are generally representative of different taxonomic groups. Therefore, Gram stain, as a classical method, is still widely used to differentiate bacteria, and clinical personnel and microbiologists use it as a preliminary method of characterization.

In addition, Gram stain or other staining methods provide contrast to assist in observation of the shape of the bacterial cell. Negative staining with nigrosin is also used to determine cell shape. Bacteria come in many shapes, but the most commonly observed are spherical [called cocci (plural) or coccus (singular)] and rod-shaped [called bacilli (plural) or bacillus (singular)].

Smith, A. C., & Hussey, M. A. (2005). "Gram Stain Protocols". Laboratory Protocols. **Microbe Library**. *American Society for Microbiology*. Accessed 13 October 2013. <http://microbelibrary.org/component/resource/gram-stain/2886-gram-stain-protocols>

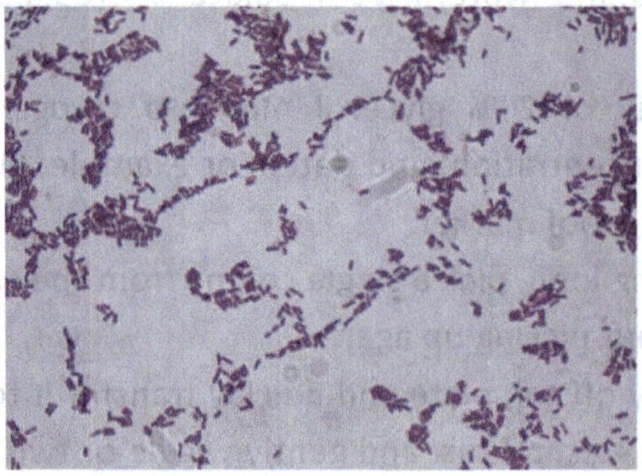

Figure 5a. *Bacillus subtilis*, a Gram-positive bacterium, stains purple with crystal violet. Photo from commons.wikimedia.org

Figure 5b. *Citrobacter freundii*, a Gram-negative bacterium, stains reddish/pink with counterstain safranin. Photo from commons.wikimedia.org

MATERIALS

- Gram stain kit
- Isolate streak plate or master plate (1- to 3-day-old culture, or fresh culture stored at 4°C)
- Positive and negative controls
- Inoculating loop, stick, or toothpick
- Microscope slide
- Forceps
- Sink or tray
- Paper towel
- Bright-field microscope and immersion oil
- Optional: Hot plate or slide warmer (alternative to Bunsen burner or flame)

PROTOCOL

1. Draw a circle on a clean, dry microscope slide and label outside with bacterial isolate designation.
2. Collect a small sample of bacterial colony with the tip of a stick/toothpick or inoculating loop.
3. Make a smear of the sample within the circle using a sterile inoculating loop or stick/toothpick. Let dry. Recommended: Smear controls side by side for comparison. A common Gram-positive control is *Staphylococcus cohnii* and a common Gram-negative control is *Escherichia coli*.
4. Using forceps, fix cells by passing slide through the Bunsen burner flame 3 or 4 times, with the smear facing up. Optional: Heat fix cells by putting slide on hot plate (set to warm or low heat) for 1–2 minutes. If the slide takes more than 2 minutes to dry, it may indicate that the smear is too thick or too watery. Once the cells are heat fixed, you may proceed to staining. Note: Heat fixing a wet sample will cause cells to boil and disrupt their cell walls and will negatively affect stain results.
5. Transfer slide to a sink or a tray where stains can be rinsed. Carefully hold one end of the slide with forceps or gloved hands. If staining multiple slides, mount the slides on a wire rack or any other surface that will allow rinsing to flow through.
6. Flood smear with crystal violet for 1 minute.

7. Gently rinse slide with an indirect stream of water from the tab or a squirt bottle for about 2 seconds. Note: Slightly tilt slide and gently dispense water over the smear, not directly onto the smear. Blot the edges of the slide on paper towel – do not wipe the smear.
8. Flood smear with Gram iodine. Wait 1 minute.
9. As in step 7, gently rinse with water and blot edges of the slide on paper towel.
10. Drip 95% alcohol solution across the smear – do this for less than 15 seconds, or until the solution runs clear off the smear. Do a quick rinse with water. Quickly blot on paper towel. Note: This is a critical step in Gram staining: Too much decolorizing agent for too long will lead to erroneous results. Note: You may use 75% ethanol and 25% acetone as an alternative decolorizer.
11. Flood smear with safranin for 1 minute.
12. Gently rinse with water and blot.
13. Let slide air-dry for several minutes.
14. View slide under the microscope under bright field. Use oil immersion for greater magnification.

MacConkey Agar Test

MacConkey agar is a specialized medium utilized to differentiate among Gram-negative bacteria, especially those that inhabit the human gastrointestinal (GI) tract. This medium inhibits the growth of Gram-positive bacteria by disrupting their cell walls with crystal violet and bile salts infused in the medium. Many Gram-negative bacteria such as *Escherichia coli* thrive in the GI tract as part of our natural gut flora, as well as many specialized Gram-positive bacteria. Other Gram-negative bacteria such as *Salmonella* and *Shigella* are notorious pathogens that can cause food poisoning and even death. To differentiate among these bacteria, MacConkey agar allows us to deduce information about their ability to ferment carbohydrates. A pH indicator in the medium changes the appearance of cells producing acid, as a by-product of fermentation, from their original color to pink/red. *Escherichia coli* ferments the lactose present in the medium and releases acid, making colonies and the surrounding agar appear red in the presence of a pH indicator. Conversely, *Salmonella* and *Shigella* do not ferment lactose and maintain a neutral pH, making colonies display their typical white or tan color (Wessner, Dupont, & Charles, p. 176). MacConkey agar is therefore both selective and differential, telling us about cell wall composition and cell's ability to ferment lactose.

REFERENCES

Wessner, D. R., Dupont, C., & Charles, T. (2013) *Microbiology* New York, NY: Wiley & Sons.

Differentiating Bacteria with the MacConkey Agar Test

Adapted from: Mary E. Allen – Microbe Library

<http://microbelibrary.org/component/resource/laboratory-test/2855-macconkey-agar-plates-protocols>

MacConkey agar is a selective and differential medium used for the isolation and differentiation of nonfastidious Gram-negative rods, particularly members of the family *Enterobacteriaceae* and the genus *Pseudomonas*. The inclusion of crystal violet and bile salts in the media prevent the growth of Gram-positive bacteria and fastidious Gram-negative bacteria, such as *Neisseria* and *Pasteurella*. The tolerance of Gram-negative enteric bacteria to bile is partly a result of the relatively bile-resistant outer membrane, which hides the bile-sensitive cytoplasmic membrane.

MACCONKEY AGAR RECIPE

Peptone (Difco) or Gelysate (BBL)	17 g
Proteose peptone (Difco) or Polypeptone (BBL)	3.0 g
Lactose	10 g
NaCl	5.0 g
Crystal Violet	1.0 mg

Neutral Red	30 mg
Bile Salts	1.5 g
Agar	13 g
Distilled Water	Add to make 1 L

Adjust pH to 7.1 ± 0.2. Boil to dissolve agar. Sterilize at 121°C for 15 minutes.

PROTOCOL

1. Streak a plate of MacConkey's agar with a pure or mixed culture. If using a mixed culture, use a streak plate or spread plate to achieve colony isolation. Good colony separation will ensure the best differentiation of lactose-fermenting and nonfermenting colonies of bacteria.
2. Streak plate of *Escherichia coli* and *Serratia marcescens* on MacConkey agar as controls. Both microorganisms grow on this selective medium because they are Gram-negative nonfastidious rods. Growth of *E. coli*, which ferments lactose, appears red/pink on the agar. Growth of *S. marcescens*, which does not ferment lactose, appears colorless and translucent.

Adapted from:

Allen, M. E. (2005). "MacConkey Agar Plates Protocols". Laboratory Protocols. **Microbe Library**. *American Society for Microbiology*. Accessed 13 October 2013. <http://microbelibrary.org/component/resource/laboratory-test/2855-macconkey-agar-plates-protocols>

PCR Analysis

Polymerase chain reaction (PCR) is a common technique used to **amplify** (make multiple copies of) specific regions of DNA for several applications, including sequencing and genetic analysis. PCR helps scientists overcome the problem of having too little material or an impure DNA sample to study. Developed in 1983 by the biochemist Kary Mullis, this elegant technique uses the same mechanism cells use to make copies of their genetic material during reproduction, except PCR does it without the cells themselves. PCR is carried out in a test tube containing a DNA template, primers, DNA polymerase, and reagents that stabilize the reaction in water. Each round of PCR requires three steps: (1) denaturation, (2) annealing, and (3) elongation, each of which doubles the amount of DNA template present in the reaction.

The **DNA template** can be any DNA molecule, such as a bacterial chromosome, a plasmid, or a fragment of DNA. The **primers** specify which region of DNA in the template is amplified. A primer is a short fragment of single-stranded DNA (ssDNA), which binds to (or anneals) to one strand of the DNA template and is recognized by the enzyme **DNA polymerase** as the point to start copying the DNA. In the first step of the reaction **(denaturation)**, the double-stranded DNA (dsDNA) template must be "unzipped" into two separate strands. To denature dsDNA, we must disrupt the hydrogen bonds between complementary base pairs by raising the reaction temperature to 94°C. In the second step **(annealing)**, the temperature is lowered to 58°C, allowing primers to bind at specific, complementary sites on the ssDNA template. There are typically two primers, one for each complementary strand. In the third step **(elongation)**, DNA polymerase recognizes the primer and starts synthesizing a new complementary DNA strand in the 5' to 3' direction.

Repeating this process over multiple rounds, typically around 30, enables exponential amplification of the DNA. The conditions of the PCR reactions require a highly stable enzyme that will withstand great fluctuations in temperature without compromising its structure and function or the quality of the new DNA copies. Just as dsDNA denatures at high temperatures, so do enzymes. But unlike DNA, whose strands reassociate upon cooling, most enzymes irreversibly lose their three-dimensional structure and activity at very high temperatures. In the past, scientists needed to add new enzymes in every round of the reaction, decreasing the efficiency and the reliability of PCR by building up a massive protein concentration through successive rounds of amplification. An extreme bacterium, *Thermus aquaticus*, a thermophilic or "heat-loving" bacterium isolated from the hot springs of the Yellowstone National Park, held the answer to this problem. This microorganism survives and reproduces at near-boiling temperatures, which also indicated that its enzymes were heat-resistant as well. A team was able to purify *Thermus aquaticus*' DNA polymerase, or *Taq* polymerase for short. This enzyme has an optimal temperature of 72°C, whereas most enzymes used in molecular biology work optimally at 37°C. *Taq* polymerase has become the standard enzyme used in PCR (Saiki et al., 1988), revolutionizing our ability to amplify DNA with high fidelity.

REFERENCES

Saiki, R. K., Gelfand, D. H., Stoffel, S., Scharf, S. J., Higuchi, R., Horn, G. T., Mullis, K. B., & Erlich, H. A. (1988) Primer-directed enzymatic amplification of DNA with a thermostable DNA polymerase. Science 239:487–491.

PCR Bead Method

MATERIALS

- 27F primer (stock solution: 20 µM)*
- 1492R primer (stock solution: 20 µM)*
- llustra™ PuReTaq™ Ready-To-Go™ PCR Bead and tube
- Sterile nuclease-free deionized water (ultrapure)
- Streak plate of isolate
- Micropipettor and tips
- Optional: Heat block

*Stored at –20°C

PROTOCOL

Protocol adapted from "puRe *Taq* Ready-To-Go PCR Beads" guide

1. Obtain PCR bead tubes, which contain *Taq* polymerase (heat-resistant enzyme) and other necessary reagents. Label tubes with names of isolates or DNA samples, and if possible, primers used.
2. Dispense 23 µL of ultrapure water into the PCR bead tube. The bead will start to dissolve and slightly effervesce. Optional: Use the heat block to preheat primer stocks to 80°C for 2 minutes, and plunge in ice prior to adding to reaction mix.
3. Add 1 µL of forward primer (27F). As you dispense the primer solution, insert the micropipette tip into the mix so that the small volume goes directly into the mix.
4. Add 1 µL of reverse primer (1492R) to the mix. This will bring the volume of the reaction mix to 25 µL.
5. Using a micropipette tip, carefully touch the colony on the streak plate. A small, visible dab of cells that barely fills the very end of the pipette tip will provide enough DNA template for the reaction. Optional: resuspend cells in 100 µL of 1× phosphate-buffered saline (PBS) solution, mix thoroughly by gently vortexing, and add 5 µL of the cell suspension to the mix. This will bring the reaction mix volume to 30 µL.
6. Dip pipette tip into reaction mix and gently swirl for 5–10 seconds to dislodge cells. Cap the tubes. If necessary, gently tap for a few seconds to attain a more uniform mix and collect all of the reagents at the bottom of the tube. Avoid forming bubbles.
7. Transfer tubes to thermal cycler (PCR machine).
8. Select appropriate program† to start cycling (about 2 hours).
9. Once cycling is complete, remove tubes and incubate on ice. Follow your instructor's instructions about storage, and follow up protocols to quality test the PCR products and prepare them for sequencing.

Protocol adapted from "puRe Taq Ready-To-Go PCR Beads" guide

†PCR cycling program:

94°C for 10 minutes – breaking down cells/ denaturation

94°C for 30 seconds – denaturation

58°C 30 seconds – annealing

72°C 1 min 50 sec (1 minute per kb of DNA template) – elongation

Cycle 30 times

72°C for 10 minutes

Reagent Method

MATERIALS

- Master mix (purchased from Promega, NEB, etc) = primers, buffer, dNTPs, $MgCl_2$, H_2O, and DNA polymerase or reagents purchased separately.
    - 0.1–0.5 µM (final concentration) forward primer
    - 0.1–0.5 µM (final concentration) reverse primer
    - Polymerase buffer (1× final concentration) (may be provided with polymerase)
    - 0.2 mM each dNTP (final concentration)
    - 1.5–2.5 mM (final concentration) $MgCl_2$
    - DNA polymerase (e.g., *Taq*) (adjust concentration depending on product specifications)
- H_2O to 25–100 µL final volume
- Template DNA = small pin-head volume of bacterial cells from streak plate OR 5 µL of cell suspension.

Optional: Make a cell suspension by resuspending cells in 100 µL of phosphate-buffered saline (PBS) solution and mix thoroughly by vortexing gently.

PROTOCOL*

1. Obtain all PCR reagents (keep frozen). Thaw on ice. Label tubes with names of samples, and if possible, primers used. Place on ice.
2. Determine the volume of each reagent to add to your reaction(s) and calculate amounts for master mix if appropriate.
3. Add each reagent to master mix starting with water and ending with the DNA polymerase enzyme. Add everything but the template (bacterial cells or cell lysis) to your master mix; check off as added. Return *Taq* enzyme to freezer.
4. Dispense master mix into individual tubes.
5. Using a micropipette tip, carefully touch the colony on the streak plate. A small dab that collects a small yet visible blob of cells will provide enough DNA template for the reaction. Or add 5 µL of the cell suspension to the mix.

6. Using a micropipettor, mix the contents of each tube by gently swishing the solution up and down several times. Cap the tubes. If necessary, gently flick or vortex for a few seconds to attain a more uniform mix.
7. Transfer tubes to thermal cycler (PCR machine).
8. Select appropriate program† to start cycling (about 2 hours).
9. Once cycling is complete, remove tubes and keep in ice. Follow your instructor's instructions about storage, and follow-up protocols to quality test the PCR products and prepare them for sequencing.

†PCR cycling program:

94°C for 10 minutes – breaking down cells/ denaturation

94°C for 30 seconds – denaturation

58°C 30 seconds – annealing

72°C 1 min 50 sec (1 minute per kb of DNA template) – elongation

Cycle 30 times

72°C for 10 minutes

Picking and Patching Colonies

Isolating single species of bacteria from a mixed culture containing tens or hundreds of species is a routine procedure that allows microbiologists to carefully examine an organism and its unique characteristics. Dilution plating allows us to spread individual bacteria on a plate to grow into distinct colonies. Yet colonies overgrow with time and mix with one another, cells migrate, and cultures get contaminated, so sterile conditions in your work environment and proper techniques are very important to isolating a bacterium and starting a pure culture.

The isolation approach we will use is aptly called "pick and patch" and involves precisely those things: "picking" bacteria from a mixed culture (e.g., usually a dilution plate) and "patching" them onto a fresh plate, which may be a pure culture or a "master plate" containing all your unique bacteria of interest for your study. The light touch of a colony with a sterile toothpick or a metal rod picks up thousands of bacteria that can be transferred and smeared or patched onto a fresh plate. At the end of this procedure you will have a "master plate" that will serve as a bacterial catalog for your experiments.

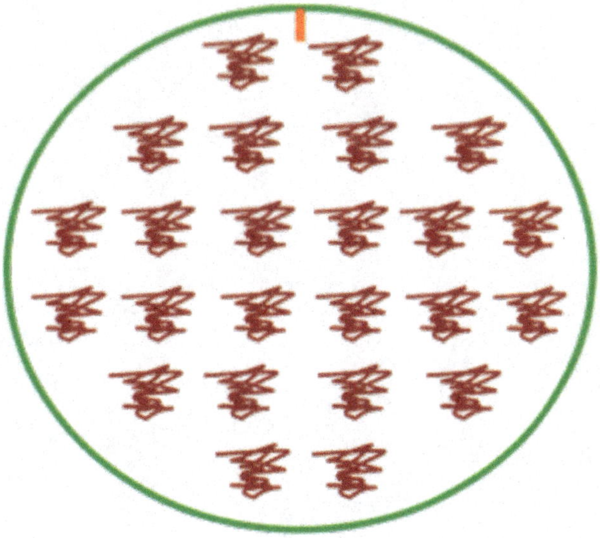

Figure 6. Master plate. This example shows how your plate should look after incubating to allow the isolates to grow. Patches (in red) were oriented on the plate using a 6×6 grid, although smaller ones can be used. Grid was aligned with point of orientation drawn on the back plate (orange vertical line at the top). Notice how patches do not touch – this is critical to prevent cross-contamination.

MATERIALS

- 2 media plates of choice
- Square grid (alternatively, draw grid on back of plates)
- Sterile toothpicks

PROTOCOL

1. Obtain 2 plates containing your media of choice. Label with date, media type, student and soil sample name around the back edge of the plate, and draw a small vertical line at the edge of your plate as point of orientation.
2. Tape the plate face-up on a square grid plate. Align grid with the vertical line on the plate.
3. Using a sterile toothpick, pick a unique colony from your dilution plate (10^{-2} or 10^{-3}). Patch (gently zigzag) colony smear on toothpick onto fresh media plate within the boundaries of one square on square grid. Be careful not to puncture the agar as you smear.
4. Continue to pick and patch colonies onto media plate, each time occupying a new square. Make sure patches do not overlap or touch because this will contaminate the patch, assuming that you picked a single colony. Try to find as many morphologically unique colonies (i.e., different textures, colony margins, pigments), but do not patch more than 24 colonies per plate to avoid overcrowding.
5. This collection of colony patches will be your master plate.

Screen for Isolate Antibiotic Production #1 – Patch/Patch

Alexander Fleming is credited with having discovered the first antibiotic introduced into widespread clinical use, penicillin. In 1928, Fleming made an observation that became the basis for the way microbiologists have searched systematically for antibiotic producing microorganisms ever since. He noticed that a mold (*Penicillium notatum*) that had contaminated a *Staphylococcus* culture was inhibiting the growth of the bacterial cells around it, creating a pronounced "zone of inhibition." The mold secretes its powerful chemical weapon into its surroundings where it diffuses into the broth or agar medium. Susceptible microorganisms that come in contact with this chemical are inhibited in their ability to survive or reproduce. Areas on a plate that would normally be lush with colonies or a lawn of bacteria become clear, creating visible zones of inhibition.

Microbiologists apply these basic principles when conducting activity assays – tests that determine the presence of antimicrobial compounds in a culture. Many researchers are concerned with finding compounds that specifically target bacteria, especially for those that pose a great threat to our health. Therefore, human pathogens or related bacteria (due to the risk to the research of working with the pathogens themselves) are used as test subjects in activity assays to find compounds that are "active" against them.

REFERENCES

Bibb, M. (1996) The regulation of antibiotic production in *Streptomyces coelicolor* A3(2). Microbiology 142:1335–1344.

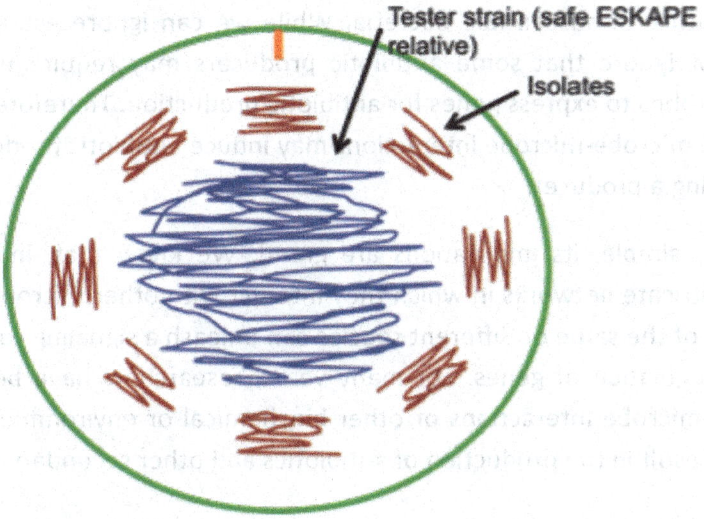

Figure 7. Patch-patch method. Candidate antibiotic producers are patched around the edge of plate, and a tester strain is patched in the middle, separated by small gap.

MATERIALS

- 2 plates containing media of choice
- Master plates (made in previous experiment) with candidate isolates to be tested

- Plate of safe ESKAPE relative as tester
- Sterile toothpicks

PROTOCOL

1. Obtain your master plate and the culture containing your safe ESKAPE relative of choice.
2. Choose a fresh medium plate to grow safe ESKAPE relative.
3. Pick isolates from your master plate and patch them around the perimeter of the plate (see diagram above)
4. Label each patch with the appropriate name on the back of the plate.
5. Patch your tester strain (safe ESKAPE pathogen) in the center of the plate near, but without touching the patches of soil isolates. Incubate at predetermined temperature for 1-3 days but no longer, to avoid overgrowth. Once patches produce dense growth, make observations or store plates at 4°C until the next lab session.

Screen for Isolate Antibiotic Production #2 – Spread/Patch

THEORY

The spread-patch protocol tests for antibiotic production by candidate isolates that are in close physical contact with the tester strain. In this protocol, the tester strain is spread on a plate, and then the isolates are patched onto the bacterial spread. If the isolate is active against the tester strain, it should theoretically have no trouble growing on the bacterial spread. Yet, this is not always the case as an antibiotic producer may need time to establish itself on the medium or may not be successful at invading a growing culture of susceptible bacteria. While we can ignore establishment times in this experiment, we cannot ignore that some antibiotic producers may require physical or biochemical contact with other microbes to express genes for antibiotic production. Therefore, this protocol is based on the assumption that microbe-microbe interactions may induce antibiotic production and will increase the chances of identifying a producer.

While this approach is simple, its implications are grand. We know that, in their natural habitats, microbes are part of intricate networks in which they interact with other microbes. Extracellular signals coming from members of the same or different species can unleash a signaling cascade within a cell that ultimately affects its regulation of genes. For many years, researchers have been attempting to find what specific microbe-microbe interactions or other biochemical or environmental cues would trigger the expression of that result in the production of antibiotics and other secondary metabolites.

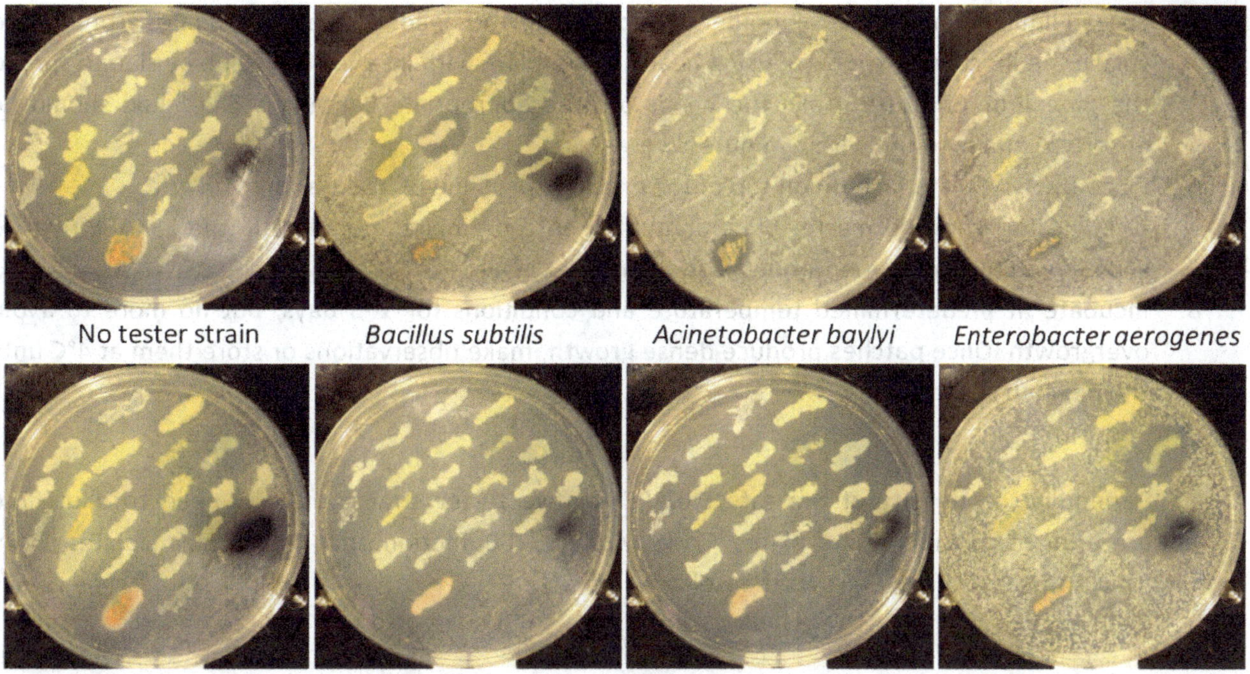

No tester strain | *Bacillus subtilis* | *Acinetobacter baylyi* | *Enterobacter aerogenes*

Erwinia carotovora | *Escherichia coli* | *Enterococcus raffinosus* | *Pseudomonas putida*

Figure 8. Spread-patch method to test isolates for antibiotic production. Tester strains were spread on PDA plates, the plates were dried, and isolates were patched onto the plate with toothpicks. Plates were incubated at 28°C for 48 h. Photo source: Manuel Fernando Garavito Diago

MATERIALS

- 2 appropriate media plates (same as master plate)
- Liquid culture of safe ESKAPE relative
- Master plates with isolates
- Spread beads or sterile swabs
- Square grid template
- Sterile toothpicks

PROTOCOL

1. Obtain appropriate media plate for your isolates to grow.
2. Label plate with respective medium, safe ESKAPE relative, and master plate used. Add vertical line as point of orientation, aligned with master plate.
3. Obtain a liquid culture of your safe ESKAPE relative.
4. Dispense 150 µL of the safe ESKAPE relative liquid culture onto the medium plate. This is the inoculation step.

4b. ***Alternatively, dip a sterile swab into ESKAPE relative and spread across plate, turning the plate 2-3 times to get even coverage of the culture on the plate.

5. Spread liquid with L-shaped spreader or add 5-10 spread beads to the plate, being careful not to splash the inoculum. If you are using beads, cover plate with lid and shake side by side to spread the inoculum. Carefully shake the beads off into appropriate container. The liquid should be absorbed into medium within minutes.
6. Place the new plate face-up on top of the grid and align line of orientation with grid.
7. Pick isolates from the master plate and patch onto the plate spread with safe ESKAPE relative arranged in the same orientation as on the master plate.
8. Incubate at predetermined temperature and conditions for 1–3 days, but no more to avoid overgrowth. Once patches produce dense growth, make observations or store them at 4°C until the next lab session.

***Depending on density of the safe ESKAPE relative tester strain you may need to dilute the culture prior to spreading onto plates. Try a 1/10 – 1/100 dilution prior to patching isolates if you find that a tester strain overgrows the isolates.

Screen for Isolate Antibiotic Production #3 – Top Agar

THEORY

Another method of screening for antibiotic production that we will be using is called top agar. In this technique, liquefied agar is mixed with the tester strain and poured over patches of the soil isolates. Since the tester strain is uniformly distributed on the overlying agar (which solidifies quickly), it could almost allow for the three-dimensional visualization of zones of inhibition.

MATERIALS

- 2 of each appropriate media plates
- Sterile glass bottle
- Liquid culture of safe ESKAPE relative
- Master plates with isolates
- Molten top agar (medium with 1/2 the amount of agar, or 0.75%, used in the typical recipe)*
- Square grid
- Sterile toothpicks

*Recommendation: Prepare top agar up to 1 day and at least 2 hours prior to lab session. Ensure medium contains a homogenous mixture of ingredients. Dispense into separate bottles (50 mL) (one per group of students working together; each student will need ~7 mL per plate) and autoclave. The agar will come out fully melted and with minimal water loss. Allow to cool (between 45°C and 55°C) until the time they will be used by the students or place in a 50-55°C water bath for use. Alternatively, microwave bottle of top agar (if made days prior or allowed to solidify) until agar melts but avoid boiling. This alternative runs the risk of water loss, resulting in higher concentrations of agar.

PROTOCOL

1. Patch isolates of interest onto appropriate solid medium for safe ESKAPE relative to grow. Do not use excessive inoculum and patch lightly.
2. Obtain bottle with molten top agar (0.75%) cooled down to 45–55°C, or comfortably warm to the touch. If too hot, top agar may kill cells in the next step. If too cold, top agar may solidify in the bottle.
3. In a sterile glass bottle, mix 100 µL of overnight liquid culture of your safe ESKAPE relative with 7 mL of top agar. Cap bottle and mix contents by swirling.
4. Gently pour top agar with safe ESKAPE relative over the patched plate. Pour onto one side of the plate and tip over to flood entire plate.
5. Cover the entire surface of the plate and let sit until top agar solidifies, at least 10–15 minutes. Incubate upside down (but be sure not to turn the plate until top agar is completely solid and rigid). Incubate under predetermined temperature and conditions. Incubate for 1–3 days, but no more to avoid overgrowth. Once isolates and tester strains produce dense growth, make observations or store them at 4°C until the next lab session.

Safe ESKAPE Relatives Information Sheet

ESKAPE Pathogen	Safe Relative	Culture media	Conditions*	ATCC #
Enterococcus faecium	*Enterococcus raffinosus*	BHI/ Sheep blood agar	30°C or 37°C	49464
	Bacillus subtilis	LB	30°C or 37°C aerobic	
Staphylococcus aureus	*Staphylococcus epidermidis*	BHI, TSA	30°C or 37°C aerobic	14990
Klebsiella species	*Escherichia coli*	NB	30°C or 37°C aerobic	11775
Acinetobacter baumannii	*Acinetobacter baylyi*	BHI	30°C or 37°C aerobic	33305
Pseudomonas aeruginosa	*Pseudomonas putida*	LB	30°C or 37°C aerobic	

ESKAPE Pathogen	Safe Relative	Culture media	Conditions*	ATCC #
Enterobacter species	*Enterobacter aerogenes*	TSA	30°C or 37°C aerobic	51697
	Erwinia carotovora	LB	30°C	
A positive control for antibiotic producers	*Lysobacter antibioticus* (10TSA2)	NB/TSA	30°C aerobic or room temp	

*For liquid culture, aerobic means incubate with shaking. *E. carotovora* and *E. raffinosus* should not be shaken.

For solid culture, all strains should be grown in regular aerobic conditions.

Gray boxes indicate Handelsman Lab, University of Wisconsin-Madison, strains.

- *E. carotovora* grows fairly slowly in liquid culture. Inoculate liquid culture 2 days in advance.
- *L. antibioticus* grows slowly on a plate. Streak out culture 2–3 days in advance.
- *S. epidermidis* grows very slowly in the antibiotic assays. Some dislike using it for that reason.
- All other strains will grow overnight without any problems. For best results, use liquid cultures within 16–24 hours after inoculation.
- *A. baylyi* produces two colony morphologies – opaque and white-colored. Both colony types have been sequenced and BLAST results indicate that both types are most closely related to *A. baylyi*. Sequence and chromatogram data are available upon request.

Review of Isolates and Activities

What media did you use?

How many colonies were picked for your master plate?

How many of those grew?

What tester strain did you use the first time?

How many active isolates did you have with this tester strain?

What two tester strains did you use the second time?

How many active isolates did you have with these tester strains?

How many active isolates did you streak out?

Table of Isolate Activity Spectrum

Overall spectrum of activity. Fill in isolate name on the top row (you can fill in up to eight isolates in this table, but feel free to reproduce the table or expand it if you have more isolates). Tester strains(safe ESKAPE relatives) appear listed in the first column. For isolates showing activity (zone of inhibition) against a tester strain, write in a plus sign (+); for no activity (no zone of inhibition), write in a minus sign (–):

Isolate name								
E. carotovora								
E. coli								
E. raffinosus								
E. aerogenes								
B. subtilis								
P. putida								
A. baylyi								
S. epidermidis								

Serial Dilutions

Adapted from: Jackie Reynolds – Microbe Library
<http://microbelibrary.org/component/resource/laboratory-test/2884-serial-dilution-protocols>

Determining microbial counts for liquid and solid samples is a common practice in the lab to quantify the biomass of a soil sample, calculate an antibiotic's minimal inhibitory concentration (MIC), or determine the population density in a liquid culture. In most environmental samples, bacteria are numerous, ranging from the tens of thousands to the millions in as little as 1 mL of seawater or 1 g of soil. Given the size of these populations and rate of cell turnover, it would be nearly impossible to get an exact count by counting all the cells directly. Instead of counting cells one by one, microbiologists calculate cell density through colony forming units (CFUs), which give us an approximation of the number of viable cells per milliliter or gram of a sample.

To calculate CFUs, a sample must be diluted in water or a saline solution that keeps the cells in suspension alive. Diluting 1 g of soil with saline solution to a final volume of 10 mL would create a 10-fold or 1:10 dilution of our soil sample; therefore, if the cells are properly suspended in the solution, all the cells contained in 1 g of sample will be evenly distributed in 10 mL of solution. Making serial dilutions of a sample in 10-fold increments allows us to reduce the number of cells per volume to a cell density that is easier to count. Once we have reached a desired dilution of our sample, we can add the dilution to a solid medium that will support the growth of the bacteria. Once the bacteria grow to colonies (1 bacterium giving rise to 1 colony of clones), we can determine how many bacteria were plated and calculate the cell density in the original sample measured in CFUs. For example, if we serially dilute 1 g of soil sample by a factor of 10^3, spread and incubate the dilution on a solid medium, and then observe 130 colonies, we would calculate that there were 130×10^3 or 1.3×10^5 CFUs/g of soil. This number represents the number of viable cells, i.e., cells in an environmental sample that can survive lab conditions and grow in culture. While the proportion of all soil bacteria we can successfully grow in the lab remains low, plating various dilutions on different types of media and under different conditions (e.g., lighting and temperature) can increase our recovery of diverse bacteria.

How to calculate CFU/g soil:

colonies ÷ volume plated (μL) × dilution factor x 1000 μL/mL × volume of suspended soil

For example:

220 colonies on 10^{-3} plate (dilution factor = 1000)

220 colonies counted ÷ 100 μL plated × 1000 × 1000 μL/mL × 10 mL/g

= 22,000,000 = 2.2×10^7 CFU/g

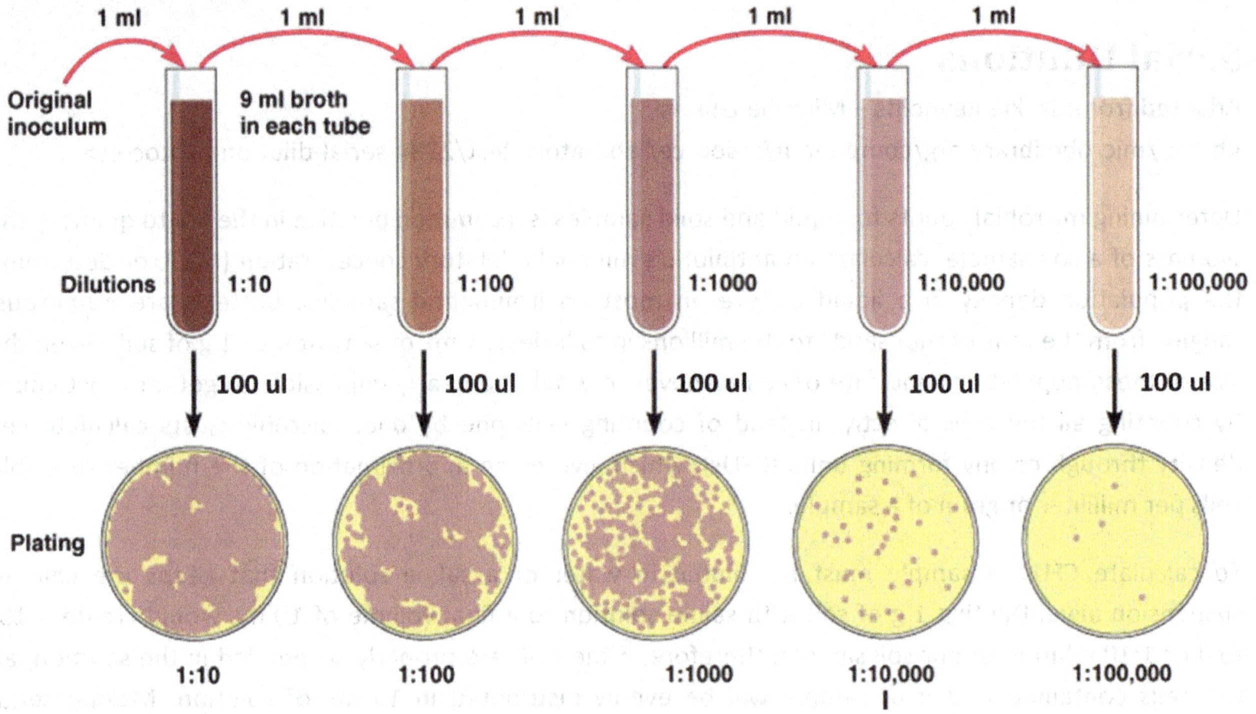

Figure 9. Serial dilution and plating schematic. A series of 10-fold dilutions is made from the original inoculum, which contains cells in suspension from the environmental sample. Each subsequent dilution plate will have 10-fold fewer bacteria than the previous one, making it possible to count individual colonies. To calculate CFU in the original inoculum, be sure to account for the effect of plating only 100 µL of a 10-mL sample. http://faculty.irsc.edu/FACULTY/TFischer/micro/serial%20dilution.jpg

MATERIALS

- Choice of media plates*
- Conical tubes or test tubes
- Soil sample
- Sterile water
- Sonicator or vortex mixer

*Refer to Media Menu at the end of this protocol.

PROTOCOL

1. Obtain and label appropriate number of plates and 1.5 mL microcentrifuge tubes (one for each subsequent dilution). Dilutions should be made in increments of 10 (10^{-1}, 10^{-2}, 10^{-3}, etc.).
2. Weigh 1 g of soil sample.
3. Transfer to test tube or 15-mL conical tube.
4. Add 9 mL of water to 1 g of soil.
5. Place tube in sonicator bath and sonicate for 30–60 sec, or vortex.

6. Determine the dilution series and calculate appropriate volumes for each. Again, dilutions should be made in increments of 10, thus add 900 µL of diluent (water) into each dilution tube. 900 µL diluent + 100 µL specimen transferred = 1000 µL
7. Remove 100 µL of soil+water with micropipette from 15-mL conical tube and add to 900 µL of water in Eppendorf tube. This is 10^{-1} dilution.
8. Mix with vortexer.
9. Remove 100 µL of 10^{-1} dilution and add to 900 µL of water in another microcentrifuge tube. Mix by vortexing.
10. Continue to transfer 100 µL of previous dilution to 900 µL of diluent until reach desired dilution.
11. When finished with dilution series, plate 100 µL of each dilution to appropriate plates.*
12. Note volume and dilutions plated.

*Check with your instructor about spread plating technique.

SAFETY

Tubes and agar plates should be discarded properly in a biohazard container for proper sterilization. The pipettes will also be sterilized (washed first if using reusable glass pipettes). Do not pipette by mouth. Use aseptic technique in the transfer of microorganisms from tube to tube, as well as in the production of the pour plates. The ASM advocates that students must successfully demonstrate the ability to explain and practice safe laboratory techniques. For more information, read the laboratory safety section of the ASM Curriculum Recommendations: Introductory Course in Microbiology and the Guidelines for Biosafety in Teaching Laboratories.

COMMENTS AND TIPS

Greater than 300 colonies on the agar plate and less than 30 leads to a high degree of error. Air contaminants can contribute significantly to a really low count. A high count can be confounded by error in counting too many small colonies or difficulty in counting overlapping colonies. Use sterile pipettes for the dilutions and use different pipettes for each dilution. To do otherwise will increase the chances of inaccurate estimates because of the carryover of cells. Accuracy in quantitation is determined by accurate pipette use and adequate mixing of the diluted samples.

Adapted from:

Reynolds, J. (2005). "Serial Dilution Protocols". Laboratory Protocols. **Microbe Library**. *American Society for Microbiology*. Accessed 17 Oct 2013. http://microbelibrary.org/component/resource/laboratory-test/2884-serial-dilution-protocols>

[redacted]

Serial Dilution Observations and Results Sheet

Pick 1–2 dilution plates from your initial serial dilution on LB and 1–2 dilution plates of your media of choice. Specify the medium in the plate and period of incubation when colonies were counted; e.g., LB at 24 hours. Express your dilution factor (usually a factor of 10) and CFU/g of soil in scientific notation (e.g., 9.7×10^7). Fill and expand the table a below as necessary.

Medium / Incubation period	Dilution factor	Volume of dilution plated	Final dilution on plate	Number of colonies	Count Cells per g of soil (CFU/g)

[redacted]

CFU Practice Exercises and Serial Dilution Questions

1. You are given a test tube containing 10 mL of a solution with 8.4×10^7 cells/mL. You are to produce a solution that contains less than 100 cells/mL. What dilutions must you perform in order to arrive at the desired result?

2. You have a microtube containing 1 mL of a solution with 4.3×10^4 cells/mL and you are to produce a solution that contains 43 cells/mL. What dilutions must you perform?

3. You are given a container with 5 mL of a solution containing 5.1×10^3 cells/mL. You are to produce a solution that contains approximately 100 cells/mL.

4. You are given the task to quantify the bacteria in your soil sample. You do a 10-fold serial dilution of your soil sample in sterile water and plate 1/10 of each subsequent (serial) dilution in LB agar plates with cycloheximide – an antifungal and eukaryotic cell inhibitor. If your 10^{-6} (ten to the minus sixth) dilution plate contains 97 bacterial colonies after two days of incubation, how many bacteria do you estimate were present in your initial soil sample (1 g of soil)? Show your calculation and briefly describe each step.

5. Notice this is a three- part question. For each answer, specify the number in the answer box. No need to restate the scenario.

Scenario: If you are determining the CFU of a sample of ocean water, how would the following scenarios (1–3) affect your CFU/mL estimate and the diversity of bacteria you observe:

1) The cells in the sample stick together in clumps

2) Most bacteria require high salt and minimal nutrients but you plate them on LB

3) You make your calculations entirely on 10-fold dilutions, but in your first dilution you put 10 µL of seawater in 1 mL of water (pay close attention to the units)

6. Is CFU/g of soil a diversity estimate or an abundance estimate? How is a diversity estimate different from an abundance estimate?

CFU Practice Exercises and Serial Dilution Answers:

1) ANSWER: You should perform a series of three 1:100 dilutions to yield 84 cells/mL.

1 mL of original solution to 99 mL of water = 8.4×10^5 cells/mL.

1 mL of second solution to 99 mL of water = 8.4×10^3 cells/mL.

1 mL of third solution to 99 mL of water = 8.4×10^1 or 84 cells/mL.

2) ANSWER: You could perform the following dilutions:

10 µL of original solution to 990 µL of water = 4.3×10^2 cells/mL.

100 µL of second solution to 900 µL of water = 4.3×10^1 or 43 cells/mL.

3) ANSWER: You would perform the following dilutions:

0.5 mL of original solution to 4.5 mL of water = 5.1×10^2 cells/mL

1 mL of second solution to 4 mL of water = 1.02×10^2 cells/mL or 10^2 cells/mL

4) ANSWER:

colonies / dilution factor = CFU/g of soil

97 colonies / 10^{-6} = 9.7×10^7 CFU/g of soil

5) ANSWER:

1) You will likely underestimate CFU/mL estimate; colonies may contain various genetically distinct cell types; colonies will merge; cells may compete for resources and grow disproportionately; you would not get a reliable determination of diversity.

2) You will get a low CFU/mL estimate (assuming that these cells prefer to grow on high salt, minimal media versus rich media like LB). In addition, fast-growing cells may obscure growth or visualization of those cells preferring minimal nutrients (which are likely to grow more slowly).

3) You have been consistent with the 1:10 ratio; you will underestimate your CFU estimate by a factor of 10; you may not get a good determination of diversity in your sample.

6) ANSWER:

Diversity – phylogenetically/genetically distinct bacteria, which may exhibit different morphologies, growth rates, metabolites, activity; different species/genera/other classifications; for our purposes – a more qualitative way to look at a sample.

Abundance = number of bacteria, usually determined by dilution plating and calculating CFU/g of sample

Silica Column Chromatography

In another section, Thin Layer Chromatography (TLC), a diagnostic tool that provides information about the polarity of an extract's active components is described. In this section, you will use the information gathered from the TLC process to guide silica column chromatography, a technique that separates organic compounds into various fractions based on polarity.

Like TLC, silica column chromatography requires a solvent system. However, the choice of solvents for the column is somewhat more complicated; rather than using just a single mixture of solvents as in the TLC protocol, you will now use a progression of increasingly polar solvent gradients. Luckily, the optimal solvent found for TLC can be used to determine the starting point of this progression. In general, it is best to begin the process with a mixture of solvents that is slightly less polar that the optimal TLC mixture. For instance, if a 4:1 hexane:ethyl acetate system proved effective in separating compounds on a TLC plate, you might begin silica column chromatography with a 6:1 hexane:ethyl acetate mixture. From this starting point, you will progress to more polar gradients. For example, a progression of gradients might begin with 6:1 hexane:ethyl acetate and continue with 4:1, 2:1, 1:1, 1:2, 1:4, 1:6, and finally 100% ethyl acetate.

MATERIALS

- 1 glass Pasteur pipette (5 ¾")
- Three-arm stand
- Wooden stick
- Cotton
- Small filter paper
- Silica gel (230–400 mesh)
- Sand
- Pipette bulb
- Solvents making up the optimal solvent system found using the "Thin Layer Chromatography (TLC)" protocol.
- Organic extract, suspended in the less polar of the chosen solvents
- 8–12 small glass test tubes (numbered)

PROTOCOL

1. Determine the progression of solvent gradients that will be used. This process is discussed above. Your system should have 5–10 different gradients.
2. Secure the Pasteur pipette on the three-arm stand with the tip pointing down.
3. Break off a small piece of cotton and insert it into the top of the pipette. Use the wooden stick to push the cotton down into the top of the narrow tip of the pipette and tamp it down gently. The cotton should be secure enough to stop the escape of silica from the column, but should not be large enough or tight enough to prevent the flow of solvents through the pipette.
4. Fold the filter paper into a small funnel. Use the funnel to pour silica gel into the top of the pipette. The level of the silica should be halfway in between the cotton plug and the indent near the top of the pipette. Tap the pipette gently until the surface of the silica settles flat.

5. Use the paper filter to add sand. The sand layer should be about 1 cm in height.
6. Put the bulb on the top of the pipette and squeeze slowly and gently. This will help push pockets of air from the column. (Caution: Do not release pressure on the bulb while it is still attached to the column. This sudden decrease in pressure will cause the silica and sand to be sucked up the pipette. Always remove the bulb from the pipette before releasing pressure.)
7. Wash the column with the more polar of the two solvents. For instance, if you plan to start with a 6:1 hexane:ethyl acetate gradient, wash with pure ethyl acetate. To do this, use another glass pipette to add this solvent to the top of the column. Add until the solvent reaches the indent near the top of the pipette. Push this solvent through with the bulb so that the solvent level is just above the sand. (Note: Never let the level of liquid in the column fall below the sand. Whenever pushing through solvent or extract, stop just before the liquid reaches the top of the sand.) Repeat once more, for a total of two washes.
8. Wash the column with the less polar of the two solvents. For instance, if you plan to start with a 6:1 hexane:ethyl acetate gradient, wash with pure hexane. Repeat once more, for a total of two washes.
9. Add the extract, suspended in the less polar of your two solvents. Add enough so that the level of the extract is halfway in between the top of the sand and the indent near the top of the pipette. Place your first test tube (#1) under the bottom of the pipette and use the bulb to push the extract down to just above the top of the sand.
10. Remove test tube #1 and replace with test tube #2. Add the first solvent gradient, filling up to the indent near the top of the pipette. Use the bulb to push the solvent down to just above the top of the sand.
11. Switch test tubes and add the next solvent gradient. Repeat this process until all gradients have been added and all fractions have been collected.

Analyzing fractions:

1. Perform a TLC of each fraction and of the raw extract (See "Thin Layer Chromatography" protocol). The stain patterns of the various fractions should "add up" to the stain pattern of the raw extract. That is, each of the components stained in the raw extract should also be visible in the same location on one or more of the fraction plates. (Note: It may not be possible to visualize some of the raw extract's components on the fraction plates due to a low concentration of the compound [for example, if the compound were spread across a large number of the fractions].)
2. Perform an overlay bioassay of all fractions using the procedure described in the "Analyze Extraction for Antibiotic Activity" section of the "Analyzing Organic Extracts for Antibiotic Production" protocol.

Soil Sampling

Soil harbors an abundance of microbial biodiversity, coming in different forms and different packages. Studies of soil samples across the planet show that even the most remote and extreme environments are rich in microbial life. Local and familiar soils are rich and dynamic, and remain largely uncharted in the laboratory. The goal in this protocol is to demonstrate just how "exotic" local soil environments can be, and how much we can learn and discover from them. Your task is to choose an ecosystem, and bring a soil sample to study in the laboratory. Pick a sample that encapsulates a good representation of your ecosystem of choice and that is rich in biodiversity.

MATERIALS

- Container: conical tube or sandwich bag
- Soil collection worksheet (paper or electronic)

PROTOCOL

- Define what biodiversity means to you and write down what indicates that an ecosystem or soil environment is rich in biodiversity
- Write down a set of criteria for picking a soil sample and choose location(s) you wish to study. Note: Be sure to check legal requirements in your state or country before transporting soil across borders. Many areas have laws regulating import or export of soil to protect natural resources, respect ownership of biodiversity, and movement of pathogens.
- Obtain 1 or 2 containers to collect a soil sample and a soil collection worksheet
- Go out into the environment and collect a small soil sample – pick 5–10 g of soil or roughly a handful of soil.
    - For conical tube, take the cap off and use the open tube to scoop soil directly
    - For sandwich bag, turn the bag inside out, grab a handful of soil, and wrap bag around soil
- Close your container and keep until next lab session. Try to keep cool and way from the sun.
- Fill in your soil collection worksheet with information about your soil sample: location, date and time, weather conditions, habitat, surrounding plant and animal life, and other descriptive information. For location and weather, feel free to use maps or weather app to get an estimate.
- Develop hypotheses about your soil sample. What do you speculate you will learn or discover from your soil sample? How does that match your original set of criteria?

Soil Sample Data Collection Sheet (Authored by: Kristen Butela – Seton Hill University)

Collected By:	
Date of Collection:	
Deth:	
Type of Soil:	
Temperature of Air:	
Temperature of Soil:	
Weather Conditions on Date of Collection:	
General Location:	
GPS Coordinates via Google Earth:	
Sample Site Descriptors:	
Additional Data to be Determined in Lab:	
pH of Soil	
Water Content	
Organic Content	

Additional Documentation:

- Photo of sample site
- Identify any plant species present near site
- Photo of soil in collection tube

Soil Sampling Kit Contents:

- Data sheets
- 50-mL conical tubes
- Lab marker
- "Scientific Soil Sample Collection Device" (plastic knife or spoon)
- Alcohol pads (used to disinfect collection device, thermometer, ruler in between samplings)
- Thermometer
- Site-marking flag (if return to site necessary)
- Plastic ruler

Printout of "Guide of Texture by Feel": <http://soils.usda.gov/education/resources/lessons/texture/>

Spread Plate

Adapted from: Kathryn Wise – Microbe Library

<http://microbelibrary.org/component/resource/laboratory-test/3085-preparing-spread-plates-protocols>

PURPOSE

One method of distributing bacteria evenly over the surface of an agar plate medium is commonly referred to as the spread plate method. Classically, a small volume of a bacterial suspension is spread evenly over the agar surface by using a sterile bent glass rod or glass beads as the spreading device. The goal in evenly distributing the bacterial suspension is typically to permit the growth of colonies that can be enumerated subsequently (see Serial Dilution Protocols) or sampled following incubation. Each plate is spread with a single inoculum of the bacterial suspension. An alternative approach to spreading a single-inoculum volume with a smooth device is to apply a smaller volume and tip the plate, allowing gravity to distribute the inoculum in a band or track (track method) or to allow the inoculum to dry in place (drop method). With this alternative approach, several sample dilutions can be distributed on a single-agar plate.

HISTORY

Since the development of the agar plate in Robert Koch's laboratory, several methods have been used to achieve an even distribution of bacterial growth on or in the agar. The most common methods used to achieve this type of distribution are: spread, pour, thin layer, layered, and membrane filter(Koch,.1994).

PRINCIPLES

Using the spread method a small volume of a bacterial suspension is distributed evenly over the surface of an agar plate using a smooth sterilized spreader (Koch, 1994). In the case of track plates, gravity is used to spread the inoculum down the agar in a column forming a track (Jett et al., 1997).

PROTOCOL

Agar plates:

Select and prepare an agar medium based upon the type of bacteria to be enumerated or selected. Freshly prepared plates do not work as well as dry plates as it takes longer for the inoculum to absorb into the agar. Plates may be dried by keeping them at room temperature for roughly 24 hours. Plates will dry faster in lower humidity, so placing them in a laminar flow hood will speed the drying process. Once dried, plates may be used or refrigerated in closed bags or containers until required. Refrigerated plates should be warmed to room temperature prior to use.

Inoculations:

When enumerating colony-forming units (CFUs), plates with 20–200 CFUs can be used to calculate the number of CFUs/mL of the original sample. Typically, a dilution series is prepared, often a ten-fold dilution series, using a suitable diluent such as phosphate-buffered saline.

Serial Dilution Protocols:

A convenient inoculum volume, in terms of spreading, absorption, and calculations, is 0.1 mL (100 μL). Since some bacteria rapidly attach to the agar surface, the inoculum should be spread soon after it is applied. Working from the most dilute suspension to the most concentrated is advised. Proceeding from most dilute to most concentrated makes it unnecessary to change pipette tips between the dilutions.

Spreading:

A reusable glass or metal spreader should be flame sterilized by dipping in alcohol (such as 70% isopropyl or ethanol), shaking off the excess alcohol, and igniting the residue. The spreader is then allowed to cool. The spreader is placed in contact with the inoculum on the surface of the plate and positioned to allow the inoculum to run evenly along the length of the spreader. Apply even, gentle pressure as the plate or spreader is spun or rotated. If glass beads are used, ensure they have been properly autoclaved and stored under sterile conditions. Pour 5–10 beads onto a plate either prior to the addition of inoculum and shake to evenly distribute until the liquid is absorbed into the plate. Used beads can be washed with soap and water or just rinsed with water prior to autoclaving.

The goal is to evenly distribute the inoculum and to allow it to be absorbed into the agar. The plate, or spreader, should be rotated long enough to avoid pooling along the spreader once the rotation is stopped.

Incubation:

After the spread plates have been permitted to absorb the inoculum for 10–20 minutes they may be inverted and incubated as desired.

Observe the plates before the colonies have had time to fully develop. Closely positioned colonies may be difficult to resolve as separate colonies later, so it may necessary to shorten the incubation period to avoid overgrowth. Incubation in closed humidified containers will help avoid plates drying out when working with slow-growing colonies.

Counting and Selection:

After incubation, plates are inspected. When plating a dilution series, the growth on the plates should reflect the predictable drop in CFUs/plate as illustrated in a 10-fold dilution series prepared from an overnight broth culture of *Escherichia coli*.

COMMENTS AND TIPS

- Make sure plates are sufficiently dry prior to use.
- Plates should be prepared in duplicate or triplicate.
- Do not delay in spreading the inoculum once it has been applied to the plate since some cells will rapidly attach to the agar, especially if the plate agar is dry.
- Avoid spreading the inoculum to the edge of the agar because it is more difficult to inspect and count colonies along the agar's edge.
- Once the dilution series has been made, inoculate plates within 30 minutes to minimize changes in the number of cells in each dilution due to cell division or death.
- Make sure even pressure is applied to the spreader so that fluid is evenly distributed along its length as the plate or spreader is rotated.
- Once the dilutions are made, work backward spreading the most dilute samples first.
- When making your own spreaders do not make the spreading edge too long, it should conveniently fit into the alcohol container as well as the plate. Fire-polish the end of the spreader the student will hold. Bending the glass to form a triangle rather than an "L" will help ensure that only smooth even surfaces touch the agar and minimize pooling.
- Distributing the organisms by rotating the spreader rather than the plate tends to cause more pooling of the inoculum.

REFERENCES

Jett, B. D., Hatter, K. L., Huycke, M. M., & Gilmore, M. S. (1997) Simplified agar plate method for quantifying viable bacteria. BioTechniques 23:648–650.

Koch, A. C. (1994) Growth measurement, p. 254–257. In Gerhardt, P., Murray, R. G. E., Wood, W. A., and Krieg, N. R. (ed.), *Methods for General and Molecular Bacteriology*. Washington, DC: ASM Press.

Adapted from:

Wise, K. (2006) "Preparing Spread Plates Protocols". *Laboratory Protocols.* **Microbe Library**. *American Society for Microbiology*. Accessed 10 Oct 2013.

<http://microbelibrary.org/component/resource/laboratory-test/3085-preparing-spread-plates-protocols>

Streak Plate

Adapted from: D. Sue Katz – Microbe Library

<http://microbelibrary.org/component/resource/laboratory-test/3160-the-streak-plate-protocol>

HISTORY

The modern streak plate procedure has evolved from attempts by Robert Koch and other early microbiologists to obtain pure bacterial cultures in order to study them, as described in an 1881 paper authored by Koch.

The earliest appearance of the three-sector streak pattern (called the T streak) commonly used today may be the 1961 photos published in Finegold and Sweeney (1960). An illustration of how to perform this streak is in the 1968 edition of the Manual of BBL Products and Laboratory Procedures. In addition to the T streak, the BBL Manual illustrates the three- and four-way streak methods, which are most commonly used today.

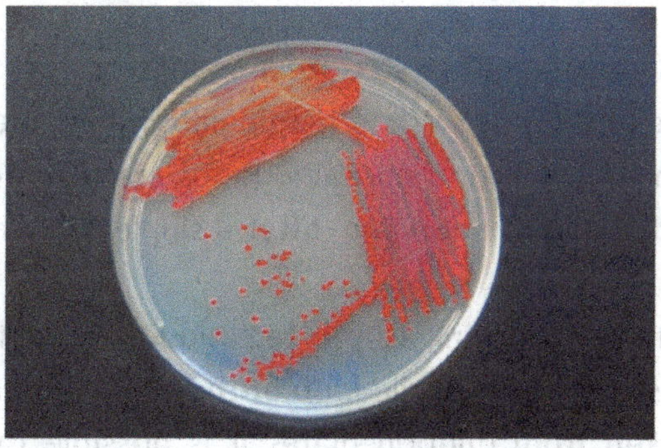

Figure 10. Four-way streak plate showing the growth of *Serratia marcescens* thinning as the streaking moves clockwise.

PURPOSE

The purpose of the streak plate is to obtain isolated colonies from an inoculum by creating areas sequentially diluting the inoculum on a single plate. Isolated colonies represent a clone of cells derived from a single precursor cell. When culture media are inoculated using a single isolated colony, the resulting culture grows from that single clone. Historically, most microbiology research and microbial characterization has been done with pure cultures.

THEORY

One bacterial cell will create a colony as it multiplies. The streak process is intended to create a region where the bacteria are so dilute that when each bacterium touches the surface of the agar, it is far enough away from other cells so that an isolated colony can develop. In this manner, spreading an inoculum containing many bacteria will result in the isolation of individual clones.

MATERIALS

- Specimen to be streaked (most likely from Master Plate)
- Wooden stick, plastic, disposable transfer loop, or reusable metal transfer loop (usually nichrome, a nickel-chromium alloy, or platinum; the single-use disposable plastic loop can be discarded between sectors rather than resterilized)
- Bunsen burner
- Sterile Petri dish with appropriate bacterial medium
- Labeling pen
- Sterile cotton swabs (if necessary to remove condensation from the agar surface and from around the inner rim of the Petri dish)

PROTOCOL

1. Label Petri dish: Petri dishes are labeled on the bottom rather than on the lid. In order to preserve an area to observe the plate after incubation, write close to the edge of the bottom of the plate. Labels usually include the organism name, type of agar, date, and the plater's name or initials.
2. Obtain sterile loop or wooden stick or sterilize a metal transfer loop by flaming in Bunsen burner so that the entire wire is red-hot: When manipulating bacteria, transfer loops, sticks, etc. are usually held like a pencil. If plastic disposable loops are being utilized, they are removed from the packaging to avoid contamination and, after a single use, are discarded into an appropriate container.
3. Open your Master Plate briefly, using aseptic technique, and collect a small number of cells from isolate patch of interest using the sterile loop or the sterile toothpick/wooden stick. Because of how patches were taken from dilution plates it is possible that each patch contains several different bacterial species or strains.
4. Streak the plate: Inoculating the agar means that the lid will have to be opened. Minimize the amount of agar and the length of time the agar is exposed to the environment during the streaking process.

   - Streak the first sector: Remove the Petri dish lid; touch the loop/toothpick on the agar in the region of the first sector and spread with bacteria by moving your loop across the dish in a zig-zag manner. Make the movements close together and cover the entire first region. The loop/toothpick should glide over the surface of the agar; take care not to dig into the agar. Discard the loop/toothpick or flame reusable loop when this is completed.
   - Between sectors: Remove the loop from the Petri dish and obtain another sterile loop/toothpick before continuing to the second sector. Either incinerate the material on the loop or obtain a sterile loop/toothpick if using disposable. Metal reusable loops must be cool before streaking can continue. Metal loops can be touched to an uninoculated area of agar to test whether they are adequately cooled. If the loop is cool, there will be no sizzling or hissing and the agar will not be melted to form a brand. If a brand is formed, avoid that area when continuing with the streaking process.

- Streak the second sector: Open the Petri dish and insert the loop. Touch the loop to the first sector once, drawing a few of the bacteria (invisible to your eye) from the first sector into the second sector. The second sector is streaked less heavily than the first sector, again using a zig-zag motion.
- Obtain a sterile loop for the third sector (see step 2, above).
- Streak the third sector: Open the Petri dish and insert the loop again. Touch the loop once into the second sector and draw bacteria from the second sector into the third sector. Streak the third sector with a zig-zag motion. This last sector has the widest gap between the rows of streaking, placing the bacteria a little further apart than in the previous two sectors. Watch closely to avoid touching the first sector as the streak is completed.
- Dispose the loop/toothpick or flame loop to sterilize for proper storage.

SAFETY

Always dispose of used loops, sticks, or any inoculating tools appropriately. Flame the metal loop one last time to sterilize for proper storage.

COMMENTS AND TIPS

- Alternative streak patterns and different culture media: A variety of alternative streak patterns are possible. Some are used for specific sources of inocula, such as urine specimens. The patterns also differ in the number of sectors as well as in the number of times the loop is sterilized. The four-quadrant streak pattern would be recommended for use when large amounts of bacteria are expected in the inoculum. The extra sector will provide additional dilution and increase the probability of isolated colonies on the plate. The four-quadrant streak plate is described in a variety of references, e.g., in Cappuccino and Sherman's *Microbiology, A Laboratory Manual*, 8th ed. (2008).
- Sometimes cultures will be streaked on enrichment media or various selective and differential media. For instance, a culture which is expected to have a Gram-negative pathogen will be streaked on a MacConkey agar plate, which inhibits the growth of Gram-positive organisms.
- Incinerating material on transfer loops—flaming: Reusable microbiological loops and needles are sterilized by flaming. A Bunsen burner is traditionally used for this process. Most microbiology manuals show the microbiologist positioned with his/her hand above the burner, with the loop placed into the flame. To avoid possible contact with the flame, the microbiologist might consider placing his/her hand below the flame with the loop/needle above the hand in the flame. The flame of the Bunsen burner should be adjusted to blue, with the darker blue cone of cooler air visible in the center of the flame. The loop or needle should be placed into the hotter part of the flame and kept there until it glows red. When flaming, the wire loop is held in the light blue area of a Bunsen burner just above the tip of inner flame of the flame until it is red hot. There is a possible aerosol hazard if the loop or needle contains liquid or a bacterial clump. Loops and needles should be placed into the heat slowly so that the moisture evaporates rather than sputters. Once sterile, the loop is allowed to cool by holding it still. Do not wave it around to cool it or blow on it.

- If an incinerator such as a Bacti-Cinerator is used to sterilize the loop, the loop is to remain inside the incinerator for 5–7 seconds. When warmed up (which will take 5 minutes), the temperature inside the incinerator is 815°C. The incinerator will take 5–10 minutes to warm up to working temperature.
- Several techniques decontaminate transfer loops between sectors of a streak plate: flame, dig into agar, flame once, and rotate loop
- There are various methods to remove organisms from the loop between sectors. Beginning students are generally taught to sterilize the loop between each sector by incinerating and then cooling the loop. Clinical microbiologists practice a variety of methods. Some flame once after the initial sector and then rotate the loop so that the next sectors can be streaked with an unused side of the loop. Other laboratorians (as clinical microbiologists call themselves) stab the loop several times into the agar to reduce the bacterial load on the loop between sectors.
- Isolated colony appearances: Isolated colonies can be described using the traditional colony descriptions. The Colony Morphology Atlas-Protocol project provides information about bacterial colony appearance and characteristic photographs. The appearance of an organism can vary. For instance, a colony of an organism growing in a crowded sector of the plate will not grow as large as the same organism growing in isolation. The media composition, pH, and moisture, as well as the incubation time and temperature can all affect the organism's appearance. Colonies selected for subculturing should be colonies that are isolated, i.e., there is no other colony visibly touching the colony.
- Agar with a surface layer of water is not suitable for obtaining isolated colonies. Obvious water drops should be removed from the surface of the plate and from the rim of the plate by using sterile cotton swabs. Plates should be incubated agar side up, to avoid condensation that would drop onto the growing colonies on the agar surface.
- Flaming tube mouths: Many protocols suggest flaming the tube mouth before and after removing organisms from a tube. Flaming is often thought to sterilize the mouth of the tube, and this was important when test tubes were capped with cotton plugs. A more important reason to flame the opening of the tube is warm the air in it, causing it expand, thereby creating a flow of air out of the tube. This reduces the chance that air will flow into the tube, possibly carrying contaminants with it. Only flame glass tubes and use extreme caution.
- Rehearsing the streak procedure: Some instructors have students practice the streaking procedure on a piece of paper. The process helps the student visualize the completed product and practice the fine muscle movements that are required in successful streaking for isolation.
- Students may also find that they can visualize the pattern better if they mark the back of the Petri dish (for instance, a T streak divide the plate into three sectors).
- Before learning to streak, students should have had the opportunity to work with 1.5% agar media. Ideally they will have had the opportunity to practice using a loop on a plate to determine the best angle of approach and the amount of force required to glide the loop over the surface of the agar without gouging the surface.
- Holding the plate while streaking: If possible, adequate lighting should be available to help the microbiologist follow the tracings of the loop on the agar. For most labs, this means that the Petri dish should be held in one's hand while being streaked in order to reflect the light properly.

Additionally, the length of time the Petri dish lid is removed should be minimized in order to limit contamination. There are several ways to hold the Petri dish. Beginning students may find that they obtain the best results leaving the plate on the lab bench and lifting the lid to work. Other students may find that they can place the plate upside down on the workbench and lift the agar-containing bottom, hold it to streak and then quickly replace it into the lid. Yet other students may have the manual dexterity to manipulate the entire dish in their hand, raising the lid with thumb and two fingers while balancing the plate in the rest of their hand.

REFERENCES

BBL. (1973) *BBL manual of products and laboratory procedures.* Cockeysville, MD: Becton Dickson Microbiology Systems.

Buchanan, E. D., & Buchanan, R. E. (1938) *Bacteriology for Students in General and Household Science* (4th ed.). New York, NY: Macmillan Company.

Cappuccino, J. G., & Sherman, N. (2008) *Microbiology a Laboratory Manual* (8th ed.). San Francisco, CA: Pearson/Benjamin Cummings.

Finegold, S. M., & Sweeney, E. E. (1960) New selective and differential medium for coagulase-positive staphylococci allowing rapid growth and strain differentiation. J Bacteriol 81(4):636–641.

Lammert, J. M. (2007) *Techniques in Microbiology. A Student Handbook.* Upper Saddle River, NJ: Pearson/Prentice Hall.

Levine, M. (1939) *An Introduction to Laboratory Technique in Bacteriology* (revised ed.). New York, NY: The Macmillan Company.

Pelczar, M. J., Jr., & Reid, R. D. (1958) *Laboratory Exercises in Microbiology,* p. 45–47. New York, NY: McGraw-Hill Book Company, Inc.

Salle, A. J. (1954) *Laboratory Manual on Fundamental Principles of Bacteriology* (4th ed.), p. 39. New York, NY: McGraw-Hill Book Company, Inc.

Sherwood, N. P., Billings, F. H., & Clawson, B. J. (1992) *Laboratory Exercises in Bacteriology and Diagnostic Methods* (7th ed.). Lawrence, KS: The World Co.

Williams, H. U. (1908) *A Manual of Bacteriology* p. 100. Revised by B. M. Bolton. Philadelphia, PA: P. Blakiston's Son & Co.

Williams, C. L., & Letton, H. P. (1916) A note on the preparation of agar agar culture media. J Bacteriol 1:547–548. http://jb.asm.org/cgi/reprint/1/5/547?maxtoshow=&HITS=10&hits=10&RESULTFORMAT=&author1=williams&author2=letton&titleabstract=agaragar&searchid=1&FIRSTINDEX=0&tdate=3/31/1931&resourcetype=HWCIT.

Adapted from:

Katz, D. S. (2008) "The Streak Plate Protocols". *Laboratory Protocols*. **Microbe Library**. *American Society for Microbiology*. Accessed 18 October 2013.
<http://microbelibrary.org/component/resource/laboratory-test/3160-the-streak-plate-protocol>

Sulfide, Indole, and Motility

Adapted from: Clifford Grimsley – Gaston College

SIM media is a semisolid differential medium, which enables detection of hydrogen sulfide, indole, and bacterial motility. Because this medium allows for the movement of bacteria, and not just growth, it will be necessary to use an inoculating needle instead of an inoculating loop to introduce the culture into the tube. Inoculating needles are essentially the same as inoculating loops; however, the wire is straight and is used to stab into the agar medium instead of spreading the bacteria on top of the agar.

Some bacteria produce hydrogen sulfide (H_2S) through the metabolism of certain amino acids containing sulfur or through the reduction of inorganic sulfur compounds in the environment, such as sulfate or sulfide. The hydrogen that is released can be detected by adding a heavy metal salt, such as lead, bismuth, or iron, into the medium. Hydrogen sulfide reacts with these compounds to produce black-colored metal sulfides.

Indole is produced when the amino acid tryptophan (yes, the one found in turkey) is metabolized. Indole production can be detected by adding a reagent known as Kovac's Reagent. If indole is produced, the Kovac's reagent will create a cherry red liquid layer on the top of the bacterial culture medium.

Bacteria have evolved several types of motility, but the most common type is driven by a motor that rotates a thick tail, called a flagellum. Motile cells can move throughout SIM medium as it is a semisolid, and they will show diffuse growth along the line of inoculation. The production of H_2S is intensified by cultures that are also motile. Therefore, if the culture is both motile and able to produce H_2S, the entire tube may turn black. Motility can also be assessed by creating a hanging drop slide of the culture and observing for motility within.

It is important to test for motility and H_2S production before testing for the presence of indole, as the test for indole production may sometimes obscure the observation of the other results.

MATERIALS

- 3 tubes of SIM media per group
- Kovac's reagent
- Cultures of test isolates and positive and negative controls
- Inoculating needle

PROTOCOL

1. Inoculate each of the cultures, respectively, into a separate tube of SIM media. Stab the wire straight down through the agar to the bottom of the tube and quickly withdraw it along the same path. Do not move the wire around in the agar.
2. Incubate the tubes at optimal growth temperature (for isolates typically 25-30°C) for 24 hours.
3. Examine the tubes for evidence of hydrogen sulfide production (browning or blackening of the media). Record the results.
4. Examine the tubes for evidence of motility of the organism. A motile species grows away from the stab line into the surrounding agar, producing lines of growth or even a general turbidity

[redacted]

throughout the tube. The growth of a non-motile organism is restricted to the path of the stab. Record your observations.

5. **Perform the Kovac test for the presence of indole, as follows:**
   a) **Pipette 2 mL of Kovac's reagent and add it to the SIM tube, being careful not to mix the reagent with the medium.**
   b) **Observe the color of the Kovac layer and record your results.**

Sulfide and Indole Production and Motility Data Sheet

Isolate name	**Sulfide**	**Indole**	**Motility**

Thin Layer Chromatography (TLC)

Using organic solvents, we can extract the active compound from antibiotic-producing isolates (see the "Analyzing Organic Extracts for Antibiotic Production" protocol). However, this extraction process is quite crude, meaning that the result of this process is not a pure active compound, but rather a mixture of many compounds. In order to further isolate the active compound, we can use a series of techniques that separate compounds based on their chemical properties. This process can be guided by the use Thin Layer Chromatography (TLC) plates.

A TLC plate is a flat glass sheet that is covered on one side with a solid absorbent (often silica). When the bottom of the plate is placed in a mixture of organic solvents, the solvents will flow through the silica and travel up the plate. If extracts are spotted onto a TLC plate, this upward solvent flow will carry with it various components of the extract. However, different compounds will flow at different rates, with nonpolar components generally traveling more quickly and polar components typically traveling more slowly. In this manner, the various compounds that make up the extract can be separated along the length of the plate, with more nonpolar molecules near the top of the plate and more polar molecules near the bottom.

The mixture of organic solvents used in this process (called the "solvent system") can also be varied to achieve different separations of the extract. For instance, the use of a more polar solvent (like methanol, for example) often carries polar compounds further up the plate, meaning that the polar compounds in the extract will be separated to a greater degree. If a more nonpolar solvent system is used, on the other hand, the result will likely be greater separation of nonpolar compounds. In general, a 1:1 mixture of hexane:ethyl acetate is a good starting point for a TLC solvent system. Other common systems include 4:1 chloroform:methanol and 1:1 dichloromethane:ethyl acetate.

Although the compounds contained in organic extracts are often white or colorless, making it difficult to identify the locations of different compounds, there are several techniques that can be used to visualize extract components and to determine which are bioactive. First, various stains can be applied to the TLC plate, resulting in the coloration of particular classes of organic compounds. In the procedure below, we recommend the use of anisaldehyde, a broad-range stain that will result in a wide range of colorations. Other stains, like ninhydrin and potassium permanganate, tend to be more specific. (For instance, ninhydrin stains amino acids and small peptides, while potassium permanganate stains oxidizable compounds.) Second, bioassay overlays can be used to determine the movement of active compounds on the plate. In this technique, a thin layer of soft agar inoculated with a bacterial tester strain is laid on top of TLC plate. After growth under the appropriate conditions, zones of inhibition can be observed, revealing the position of the active compound(s).

MATERIALS

- 2 small silica TLC plates (about 6 cm in height and 2–4 cm in length)
- Organic extract
- 50 mL beaker
- Solvents
- Micropipettes or glass capillary tube

- Black Sharpie marker (as a "control")
- Forceps
- UV lamp
- Anisaldehyde stain (prepared by teaching assistant)
- Heat dryer or hot plate
- LB top agar (0.75%)
- Liquid culture of tester strain

PROTOCOL

1. Obtain dried organic extract.
2. Re-suspend extract in 100ul of the solvent used in extraction.
3. Draw a thin pencil line on the silica side of each TLC plate about 0.75 cm from the bottom. Be gentle so as not to scratch the silica from the glass surface. In addition, label one plate to indicate which will be used for staining and which will be used for the bioassay.
4. Use micropipette or a thin glass capillary tube to spot 2ul of extract on a point along the pencil line.
5. Wait for the extract spot to dry nearly completely. Then, once again spot 2ul of extract at the same point on the pencil line. Continue this process until 10 spots have been placed on the stain plate and 20 spots have been placed on the bioassay plate.
6. Choose another point on the pencil line and gently tap tip of black Sharpie marker on plate until a small black dot is formed. Be sure to get a sufficient amount of ink on the plate, but be careful not to pull the silica off the glass surface. The marker will act a control and allow you to visualize movement up the plate with the solvents.
7. Select solvent system and mix solvents in 50ml glass beaker. The total volume of the system should be 1 mL. For instance, if the 1:1 hexane:ethyl acetate system is chosen, 500 µL of hexane and 500 µL of ethyl acetate should be added to the beaker.
8. Use forceps to gently pick up each plate from the top. Place the plate in the beaker such that the plate sits level (not crooked). The solvent system should cover the bottom of the plate, but be well below the pencil line. Cover the beaker with aluminum foil to prevent evaporation of solvents. You should be able to see solvent line moving upward across the silica.
9. Allow the solvent to run until the solvent front is about 0.5 cm from the top of the plate. At this point, use forceps to remove the plate and gently blot edges with paper towel. Allow excess solvent to evaporate completely in the fume hood.
10. Shine UV lamp on the TLC plate prepared for staining and make note of areas of the plate that fluoresce. Try not to shine UV light on plate for more than a few seconds.

For staining TLC plate:

1. Obtain the anisaldehyde stain prepared by your teaching assistant. **Note:** This stain contains sulfuric acid; handle this stain carefully. Using forceps, pick up the plate prepared for staining by its top edge. Dip entire plate into anisaldehyde and quickly remove. Plate can be placed on a Petri dish lid or other surface to prevent staining the chemical fume-hood surface.

2. Blot edges of stained plate on paper towel. Dry and heat plate using heat dryer or hot plate until background of plate is light pink and spots of various colors appear. Take pictures shortly after heating, as colors may begin to fade.

For the bio-assay plate:

1. Dampen a KimWipe with ethanol and gently wipe back and edges of the plate prepared for bioassay. Place this cleaned plate face up in an empty Petri dish.
2. Obtain a liquid culture of a bacterial tester strain of your choice. This strain should be one that was inhibited by your organic extract. Inoculate 12 mL of warm LB top-agar or soft agar for overlay (0.75%) with 240 µL of liquid culture.
3. Pour spiked soft agar into the Petri dish. Swirl gently, ensuring that agar is distributed evenly across Petri dish and that the entirety of the TLC plate is covered.
4. Incubate under appropriate conditions for the tester strain. Keep the Petri dish face up to avoid disturbing the agar.

Further Steps

Observe the results of the anisaldehyde staining and the bioassay overlay to determine how well your chosen solvent system separated the components of your extract. An ideal separation would have compounds spread over the entire length of the TLC plate rather than grouped closely together in one small section of the plate. Adjust the solvent system to try to optimize separation. In general, if the components are closely grouped at the bottom of the plate, make the solvent system more polar. For example, if a 4:1 chloroform:methanol system results in a tight group of compounds at the bottom of the plate, you might make the system more polar by adjusting the ratio to 2:1 chloroform:methanol. If, on the other hand, the components are closely grouped at the top of the plate, make the solvent system more nonpolar. If, for example, a 1:1 hexane:ethyl acetate results in tight group of compounds near the top of the plate, you might make the system less polar by adjusting the ratio to 3:1 hexane:ethyl acetate. Repeat this process of trials and readjustment until the optimal solvent system is achieved.

Typical Media Menu

This list includes common microbiological media (and their recipes). Most media come premixed in a dehydrated powder form that we mix with deionized water. Check which media types are available (ask your instructor) and select a media type to grow your soil sample – you will be using this medium for serial dilutions and to grow the your isolate from your dilution plates. Also, pay close attention to their ingredients and how they relate to bacterial nutrition. For more media types and recipes, refer to the Difco "Manual of Microbiological Culture Media" (part of BD Bioscience). <http://www.bd.com/ds/technicalCenter/misc/difcobblmanual_2nded_lowres.pdf>

Luria Broth or Lysogeny Broth (LB)

Rich media for propagation and maintenance of *E. coli* used in molecular biology.

- Tryptone – 10 g
- Yeast Extract – 5 g
- Sodium Chloride – 10 g
- Agar - 15 g

pH ~7.0

Tryptic Soy Agar (TSA) – in 1/10 or 1/2 strength

General medium used for cultivation of a wide variety of microorganisms.

- Pancreatic Digest of Casein – 17 g
- Papaic Digest of Soybean – 3 g
- Dextrose (Glucose) – 2.5 g
- Sodium Chloride – 5 g
- Dipotassium Phosphate – 2.5 g
- Agar – 15 g

pH ~7.3V V

Potato Dextrose Agar (PDA)

Cultivation of yeasts and molds

- Potato Starch – 4 g
- Dextrose (Glucose) – 20 g
- Agar – 15 g

pH ~5.1

All Culture (AC)

For cultivating a wide variety of microorganisms

- Proteose Peptone No. 3 – 20 g
- Beef Extract – 3 g
- Yeast Extract – 3 g

- Malt Extract – 3 g
- Dextrose (Glucose) – 5 g
- Ascorbic Acid – 0.2 g
- Agar - 15 g

pH ~7.2

R2A

Low nutrient (minimal) medium for enumeration and cultivation of bacteria from potable water.

- Yeast Extract – 0.5 g
- Acid Digest of Casein – 0.5 g
- Pancreatic Digest of Casein – 0.25 g
- Peptic Digest of Animal Tissue – 0.25 g
- Dextrose (Glucose) – 0.5 g
- Soluble Starch – 0.5 g
- Sodium Pyruvate – 0.5 g
- Potassium Phosphate, Dibasic – 0.3
- Magnesium Sulfate – 0.024 g
- Agar – 15 g

pH ~7.2

Brain Heart Infusion (BHI)

Cultivation of fastidious (fussy) microorganisms, including streptococci, pneumococci, and meningococci

- Brain Heart, Infusion from (Solids) – 6 g
- Peptic Digest of Animal Tissue – 6 g
- Sodium Chloride – 5 g
- Dextrose (Glucose) – 3 6
- Pancreatic Digest of Gelatin – 14.5 g
- Disodium Phosphate – 2.5 g
- Agar - 15 g

pH ~7.4

Todd Hewitt (TH)

Cultivation of group A streptococci or as a blood culture medium

- Heart, Infusion from 500 g – 3.1 g
- Neopeptone – 20.0 g
- Dextrose (Glucose) – 2 g
- Sodium Chloride – 2 g
- Disodium phosphate – 0.4 g
- Sodium carbonate – 2.5 g
- Agar – 15 g

pH ~7.8

Lab Notes

Lab Notes

Lab Notes

Lab Notes

Lab Notes

Lab Notes

Lab Notes

Lab Notes